# Lecture Notes in Mathematics

A collection of informal reports and seminars
Edited by A. Dold, Heidelberg and B. Eckmann, Zürich

28

P. E. Conner · E. E. Floyd
University of V[illegible]



# The Relation of Cobordism to K-Theories

1966

Springer-Verlag · Berlin · Heidelberg · New York

Library of Congress Catalog Card Number 66-30143. Printed in Germany. Title No. 7348.

## INTRODUCTION

These lectures treat certain topics relating K-theory and cobordism. Since new connections are in the process of being discovered by various workers, we make no attempt to be definitive but simply cover a few of our favorite topics. If there is any unified theme it is that we treat various generalizations of the Todd genus.

In Chapter I we treat the Thom isomorphism in K-theory. The families U, SU, Sp of unitary, special unitary, symplectic groups generate spectra MU, MSU, MSp of Thom spaces. In the fashion of G. W. Whitehead [26], each spectrum generates a generalized cohomology theory and a generalized homology theory. The cohomology theories are denoted by $\Omega_U^*(\cdot)$, $\Omega_{SU}^*(\cdot)$, $\Omega_{Sp}^*(\cdot)$ and are called cobordism theories; the homology theories are denoted by $\Omega_*^U(\cdot)$, $\Omega_*^{SU}(\cdot)$, $\Omega_*^{Sp}(\cdot)$ and are called bordism theories. The coefficient groups are, taking one case as an example, given by $\Omega_U^n = \Omega_U^n$ (point), $\Omega_n^U = \Omega_n^U$ (point) and are related by $\Omega_n^U = \Omega_U^{-n}$. Moreover $\Omega_n^U$ is just the bordism group of all bordism classes $[M^n]$ of closed weakly almost complex manifolds $M^n$, similarly for $\Omega_n^{SU}$ and $\Omega_n^{Sp}$. On the other hand there are the Grothendieck-Atiyah-Hirzebruch periodic cohomology theories $K^*(\cdot)$, $KO^*(\cdot)$ of K-theory. The main point of Chapter I, then, is to define natural transformations

$$\mu : \Omega_{SU}^*(\cdot) \rightarrow KO^*(\cdot)$$

$$\mu_c : \Omega_U^*(\cdot) \rightarrow K^*(\cdot)$$

of cohomology theories. Such transformations have been folk theorems since the work of Atiyah-Hirzebruch [6], Dold [13], and others. It should be noted that on the coefficient groups,

$$\mu_c : \Omega_U^{-2n} \to K^{-2n}(pt) = Z$$

is identified up to sign with the Todd genus $Td : \Omega_{2n}^U \to Z$.

In Chapter II we show among other things that the cobordism theories determine the K-theories. For example, $\mu_c$ generates a ring homomorphism $\Omega_U^* \to Z$ and makes Z into a $\Omega_U^*$-module. It is shown that

$$K^*(X,A) \approx \Omega_U^*(X,A) \otimes_{\Omega_U^*} Z$$

as $Z_2$-graded modules. Similarly symplectic cobordism determines $KO^*(\cdot)$. The isomorphisms are generated by $\mu_c, \mu$ respectively. Various topics are treated along the way, in particular cobordism characteristic classes.

There is the sphere spectrum $\mathcal{S}$, whose homology groups are the framed bordism groups $\Omega_*^{fr}(\cdot)$. The coefficient group $\Omega_n^{fr}(\text{point}) = \Omega_n^{fr}$ are just the stable stems $\pi_{n+k}(S^k)$, k large. The spectrum $\mathcal{S}$ is embedded in a natural way in MU, and one can thus form $MU/\mathcal{S}$. In Chapter III we study the group

$$\Omega_n^{U,fr} = \pi_n(MU/\mathcal{S}) = \pi_{n+2k}(MU(k)/S^{2k}),$$

k large. The elements of $\Omega_n^{U,fr}$ are interpreted as bordism classes $[M^n]$ of compact (U,fr)-manifolds $M^n$, where roughly a (U,fr)-manifold is a differentiable manifold M with a given complex structure on its stable tangent bundle $\tau$ and a given compatible framing of $\tau$ restricted to the boundary $\partial M$. These bordism classes have Chern numbers and hence a Todd genus

$$Td : \Omega^{U,fr}_{2n} \longrightarrow Q, \text{ Q the rationals.}$$

It is proved that given a compact (U,fr)-manifold $M^{2n}$, there is a closed weakly almost complex manifold having the same Chern numbers as $M^{2n}$ if and only if Td $[M^{2n}]$ is an integer; this makes use of recent theorems of Stong [23] and Hattori [15]. There is a diagram

$$\begin{array}{ccccccccc} 0 & \longrightarrow & \Omega^{U}_{2n} & \longrightarrow & \Omega^{U,fr}_{2n} & \longrightarrow & \Omega^{fr}_{2n-1} & \longrightarrow & 0 \\ & & & & \big\downarrow{\scriptstyle Td} & & & & \\ & & & & Q & & & & \end{array}$$

which gives rise to a homomorphism

$$E_U : \Omega^{fr}_{2n-1} \longrightarrow Q/Z.$$

This turns out to coincide with a well-known homomorphism of Adams,

$$e_c : \Omega^{fr}_{2n-1} \longrightarrow Q/Z.$$

We are thus able to give a cobordism interpretation of the results of Adams [13] on $e_c$. It should be pointed out that Chapter III is in large part independent of Chapter II.

It is to be noted that we have omitted spin cobordism completely; this is because of our ignorance. However the recent work of Anderson-Brown-Peterson is a notable example of the application of K-theory to cobordism.

CONTENTS

## CHAPTER I. THE THOM ISOMORPHISM IN K-THEORY.

Given a U(n)-bundle $\xi$ over a finite CW complex X there is constructed an element $\mathcal{J}(\xi) \in \widetilde{K}(M(\xi))$ where $M(\xi)$ is the Thom space of $\xi$; we call $\mathcal{J}(\xi)$ the Thom class of $\xi$. Similarly given an SU(4k)-bundle there is constructed a Thom class $t(\xi) \in \widetilde{KO}(M(\xi))$, and given an SU(4k + 2)-bundle there is constructed a class $s(\xi) \in \widetilde{KSp}(M(\xi))$. These Thom classes give rise to isomorphisms

$$K(X) \approx \widetilde{K}(M(\xi))$$
$$KO(X) \approx \widetilde{KO}(M(\xi))$$
$$KO(X) \approx \widetilde{KSp}(M(\xi))$$

in the three cases. Formulas for the Chern character $ch\,\mathcal{J}(\xi)$ are obtained.

No claims for originality are made in this chapter; the methods have been well-known since the work of Atiyah-Hirzebruch [6], Dold [13], and others [7]. However since the results are needed explicitly in the later chapters we include an exposition. A deviation from the standard treatment is made in that exterior algebra is used in all cases, thus avoiding the use of Clifford algebras.

The chapter terminates with the setting up of homomorphisms $\Omega^*_{SU}(\cdot) \to KO^*(\cdot)$ and $\Omega^*_U(\cdot) \to K^*(\cdot)$ of cohomology theories, where $\Omega^*_{SU}(\cdot)$, $\Omega^*_U(\cdot)$ denote the cohomology theories based on the spectra MSU, MU.

### 1. Exterior algebra

We fix in this section a complex inner product space V of dimension n, and we also fix a unit vector $\sigma \in \Lambda^n V$. If $n = 4k + 2$, we make the exterior algebra $\Lambda V$ into a quaternionic vector space. If

$n = 4k$ then a real vector subspace $RV$ of $\wedge V$ is selected so that $\wedge V$ is identified with the complexification of $RV$. The special unitary group $SU(n)$ operates in a quaternionic linear fashion on $\wedge V$ in the first case, in a real linear fashion on $RV$ in the second case.

Fix, then, the complex inner product space $V$ of dimension $n$. To fit with quaternionic notation, the complex numbers are taken to act on the right and the inner product $\langle , \rangle$ is taken conjugate linear in the first variable and complex linear in the second.

There is the graded exterior algebra $\wedge V = \sum_0^n \wedge^k V$ with $\wedge^0 V = C$ and $\wedge^1 V = V$. The inner product on $V$ can be extended to an inner product on $\wedge V$ by

i) if $j \neq k$ then $\wedge^j V$ is orthogonal to $\wedge^k V$,

ii) if $X = u_1 \wedge \cdots \wedge u_k$ and $Y = y_1 \wedge \cdots \wedge y_k$ where $u_r, v_s \in V$, then

$$\langle X,Y \rangle = \det | \langle u_r, v_s \rangle | .$$

If $e_1, \ldots, e_n$ is an orthonormal basis for $V$ then the $e_{r_1} \wedge \cdots \wedge e_{r_k}$ with $r_1 < \cdots < r_k$ form an orthonormal basis for $\wedge^k V$. There is also a canonical anti-isomorphism $\alpha$: $\wedge V \to \wedge V$ with

$$\alpha(v_1 \wedge \cdots \wedge v_k) = v_k \wedge \cdots \wedge v_1 = (-1)^{k(k-1)/2} v_1 \wedge \cdots \wedge v_k .$$

It is clear that $\alpha$ is unitary.

DEFINITION. By an <u>SU-structure</u> for $V$ we shall mean a unit vector $\sigma \in \wedge^n V$; suppose an SU-structure has been fixed for $V$. Define a real linear map $\tau$: $\wedge^k V \to \wedge^{n-k} V$ as follows: fix $X \in \wedge^k V$ and let $Y$ vary over $\wedge^{n-k} V$ so that $\langle \sigma, X \wedge Y \rangle$ is a linear map $\wedge^{n-k} V \to C$; define $\tau X$ to be the unique element of $\wedge^{n-k} V$ such that

$$\langle \tau X, Y \rangle = \langle \sigma, X \wedge Y \rangle , \text{ all } Y \in \wedge^{n-k} V.$$

It is then seen that the above equation holds for all $Y \in \wedge V$.

The map $\tau$ is conjugate linear. For

$$\langle \tau(Xa), Y \rangle = a \langle \sigma, X \wedge Y \rangle = \langle (\tau X)\bar{a}, Y \rangle$$

and $\tau(Xa) = (\tau X)\bar{a}$.

Fix an orthonormal basis $e_1, \cdots, e_n$ of $V$ such that the given SU-structure is $\sigma = e_1 \wedge \cdots \wedge e_n$. By a monomial of $\wedge V$ we mean an element $X = \pm e_{r_1} \wedge \cdots \wedge e_{r_k}$ where $r_1 < \cdots < r_k$. It is seen that if X and Y are monomials, then

$$\langle X, Y \rangle = \begin{cases} 1 \text{ if } Y = X \\ -1 \text{ if } Y = -X \\ 0 \text{ otherwise.} \end{cases}$$

Moreover given a monomial X there is a unique monomial $\tilde{X}$ with $X \wedge \tilde{X} =$

(1.1) If X is a monomial then $\tau X$ is the unique monomial X with $X \wedge \tilde{X} = \sigma$.

This is readily seen from the definition of $\tau$.

(1.2) We have $\tau^2 X = (-1)^{k(n-k)} X$ for $X \in \wedge^k V$.

Proof. It is sufficient to prove (1.2) for monomials. For X a monomial, $\tau X$ is the unique monomial with $X \wedge \tau X = \sigma$. Then $\tau X \wedge X = (-1)^{k(n-k)} \sigma$ and $\tau^2 X = (-1)^{k(n-k)} X$.

Define an operator $\mu : \wedge^k V \to \wedge^{n-k} V$ by $\mu = \tau\alpha$. Then $\mu$ is conjugate linear.

(1.3) We have $\mu^2 X = (-1)^{n(n-1)/2} X$ for $X \in \wedge V$.

Proof. It is seen from (1.2) that $\mu^2 X = (-1)^r X$ where

$$r = k(k-1)/2 + (n-k)(n-k-1)/2 + k(n-k)$$
$$= k(n-1)/2 + (n-k)(n-1)/2 = n(n-1)/2.$$

The remark follows.

We now identify U(n) with the group of linear maps $g : V \to V$ with $\langle gu, gv\rangle = \langle u,v\rangle$ for all u, v ε V. Then U(n) acts on $\Lambda V$ by $g(v_1 \wedge \cdots \wedge v_k) = gv_1 \wedge \cdots \wedge gv_k$. Identify the special unitary group SU(n) with the set of all g ε U(n) for which $g(\sigma) = \sigma$.

(1.4) _If_ g ε SU(n), _then_ $g\tau = \tau g$ _and_ $\mu g = g\mu$.

Proof. From $\langle \tau X,Y\rangle = \langle \sigma, X \wedge Y\rangle$ we get

$$\langle g\tau X, gY\rangle = \langle \sigma, gX \wedge gY\rangle = \langle \tau gX, gY\rangle ,$$

hence $g\tau = \tau g$. It follows immediately that $g\mu = \mu g$.

(1.5) THEOREM. _Consider the complex inner product space_ V _of dimension_ n, _with given SU-structure_ $\sigma \varepsilon \Lambda^n V$. _If_ n = 4k + 2 _then_ $\Lambda V$ _becomes a right quaternionic vector space by defining_ $Y \cdot j = \mu(Y)$ _for_ $Y \varepsilon \Lambda V$. _Moreover_ SU(n) _acts on_ $\Lambda V$ _in a quaternionic linear fashion_. _If_ n = 4k, _let_ R(V) _be all_ X ε $\Lambda V$ _with_ $\mu X = X$ _and_ $R_-(V)$ _all_ X _with_ $\mu X = -X$; _then_

$$\Lambda V = RV + R_-(V)$$

_is a splitting into real vector subspaces and multiplication by_ i _takes_ RV _into_ $R_-(V)$ _and_ $R_-(V)$ _into_ RV. _Moreover_ SU(n) _acts on_ RV _in a real linear fashion_.

Proof. Consider the case n = 4k + 2. It follows from (1.3) that $\mu^2 = -1$. Also $\mu$ is conjugate linear so that

$$Xij = \mu(Xi) = -(\mu X)i = -Xji.$$

It follows that there is defined an action of the quaternions H on $\Lambda V$, and $\Lambda V$ is a quaternionic vector space. Consider g ε SU(n). Then

$$g(Xj) = g\mu(X) = \mu g(X) = (gX)j$$

using (1.4), so that SU(n) acts in a quaternionic linear fashion. If $n = 4k$, we have $\mu^2 = 1$. Hence $\Lambda V = RV \oplus R_-(V)$. If $X \in RV$, then

$$\mu(Xi) = -(\mu X)i = -Xi$$

and $Xi \in R_-(V)$. The theorem is then proved.

Let $\Lambda^{od}V = \sum \Lambda^{2k+1}V$, $\Lambda^{ev}V = \sum \Lambda^{2k}V$; similarly define $R^{od}V$ and $R^{ev}V$. If $n = 2 \bmod 4$ then SU(n) acts on the quaternionic vector spaces $\Lambda^{od}V$ and $\Lambda^{ev}V$. If $n = 0 \bmod 4$ then SU(n) acts on the real vector spaces $R^{od}V$ and $R^{ev}V$.

## 2. Tensor products of exterior algebras.

Let V and W be complex inner product spaces of dimension m,n respectively, with given SU-structures $\sigma_1$ and $\sigma_2$. Using the identification $\Lambda(V + W) = \Lambda V \otimes \Lambda W$ of graded algebras, then V + W receives the SU-structure $\sigma = \sigma_1 \otimes \sigma_2$. According to section 1, if $m = 2 \bmod 4$ we consider $\Lambda V$ as a $Z_2$-graded quaternionic vector space while if $m = 0 \bmod 4$ we obtain a $Z_2$-graded real vector space RV. A main purpose of this section is to prove the following.

(2.1) THEOREM. <u>There exist natural isomorphisms</u>

$$R(V + W) \approx R(V) \otimes_R R(W), \quad m = 4k,\ n = 4\ell$$
$$\Lambda(V + W) \approx R(V) \otimes_R \Lambda(W), \quad m = 4k,\ n = 4\ell + 2$$
$$\Lambda(V + W) \approx \Lambda(V) \otimes_R R(W), \quad m = 4k + 2,\ n = 4\ell$$
$$R(V + W) \approx \Lambda(V) \otimes_H \Lambda(W), \quad m = 4k + 2,\ n = 4\ell + 2.$$

In cases 1 and 4, the vector spaces and the isomorphisms are taken to be real linear, while in cases 2 and 3 they are taken to be quaternionic linear.

For each $v \neq 0$ in V we also obtain isomorphisms

$$\varphi_V : \Lambda^{od}V \approx \Lambda^{ev}V, \quad m = 4k + 2$$
$$\varphi_V : R^{od}V \approx R^{ev}V, \quad m = 4k.$$

The proof of (2.1) is based on the following lemma.

(2.2) LEMMA. *If* $X \in \Lambda^r V$ *and* $Y \in \Lambda^s W$ *then*

$$\mu(X \otimes Y) = (-1)^{ms}\, \mu_1(X) \otimes \mu_2(Y)$$

*where* $\mu, \mu_1, \mu_2$ *denote the maps of section 1 for* $\Lambda(V + W) = \Lambda V \otimes \Lambda W$, $\Lambda V$, $\Lambda W$ *respectively.*

Proof. Fix an orthonormal basis $e_1, \cdots, e_m$ for V and $e_{m+1}, \cdots, e_{m+n}$ for W such that

$$e_1 \wedge \cdots \wedge e_m = \sigma_1, e_{m+1} \wedge \cdots \wedge e_{m+n} = \sigma_2.$$

By (1.1), $\tau_1 X$ is the unique monomial $\tilde{X}$ with $X \wedge \tilde{X} = \sigma_1$ and $\tau_2 Y$ is the unique monomial Y with $Y \wedge \tilde{Y} = \sigma_2$. Then

$$(X \wedge \tilde{X}) \otimes (Y \wedge Y) = \sigma$$

$$(-1)^{s(m-r)}(X \otimes Y) \wedge (\tilde{X} \otimes \tilde{Y}) = \sigma$$

and $\tau(X \otimes Y) = (-1)^{s(m-r)}\ \tau_1 X \otimes \tau_2 Y$. Since $\alpha(X \otimes Y) = (-1)^{rs}\ \alpha X \otimes \alpha Y$, then

$$\tau\alpha(X \otimes Y) = (-1)^{ms}\ \tau_1 \alpha X \otimes \tau_2 \alpha Y$$

and the result follows. Note that if m is even then $\mu = \mu_1 \otimes \mu_2$.

*We consider now the proof of* (2.1) *for* $m = 4k$ and $n = 4\ell$. There is a natural homomorphism

$$\gamma : \Lambda V \otimes_R \Lambda W \rightarrow \Lambda V \otimes_C \Lambda W$$

whose kernel is generated by all $Xi \otimes Y - X \otimes Yi$. On the real tensor product there is the involution $\mu_1 \otimes \mu_2$, and among its fixed vectors there is $RV \otimes_R RW$. Consider then

$$RV \otimes_R RW \longrightarrow \Lambda V \otimes_C \Lambda W = \Lambda(V + W)$$

which by (2.1) has image in $R(V + W)$. It is seen that if $y \in$ Kernel $\gamma$, then

$$(1 \otimes i)y = -(i \otimes 1)y.$$

If also $y \in R(V) \otimes_R R(W)$ then the left hand side belongs to $R_-(V) \otimes_R R(W)$ and the right hand side to $R(V) \otimes_R R_-(W)$ by (1.5). Hence $y = 0$, and $RV \otimes_R RW$ maps monomorphically into $R(V + W)$. Since the two are seen to have the same dimension, then

$$RV \otimes_R RW \approx R(V + W).$$

It is also seen that the actions of $SU(m) \times SU(n)$ on the two sides are identified.

If $m = 4k$ and $n = 4\ell + 2$ then one sets up similarly an isomorphism $RV \otimes_R \Lambda W \approx \Lambda(V + W)$ of quaternionic vector spaces, where $q \in H$ acts on the left hand side by $1 \otimes q$.

<u>Consider finally the case</u> $m = 4k + 2$, $n = 4\ell + 2$. Define a left action of $H$ on $\Lambda W$ by $q \cdot Y = Y \cdot \bar{q}$, so that we obtain a real vector space $\Lambda V \otimes_H \Lambda W$. Here we write an element $q$ as $\alpha + \beta j$ where $\alpha, \beta \in C$ and define $\bar{q} = \alpha - \bar{\beta} j$; this is an anti-automorphism of $H$. There is a natural epimorphism

$$\gamma' : \Lambda V \otimes_C \Lambda W \longrightarrow \Lambda V \otimes_H \Lambda W.$$

If $X \in \Lambda V$ and $Y \in \Lambda W$, then $\gamma'$ maps $\mu(X \otimes_C Y)$ and $X \otimes_C Y$ into the same value. For we have

$$\mu(X \otimes_C Y) = \mu_1 X \otimes_C \mu_2 Y = Xj \otimes_C Yj = -(Xj \otimes_C jY).$$

But $-(Xj \otimes_H jY) = X \otimes_H Y$. It is thus seen that

$$\text{Kernel } \gamma' \supset R_-(V + W).$$

A check of dimensions reveals that we have Kernel $\gamma' = R_-(V + W)$, since $\gamma'$ is an epimorphism. Hence

$$\gamma' : R(V + W) \approx \Lambda V \otimes_H \Lambda W,$$

and (2.1) is proved.

Return now to a single complex inner product space V of finite dimension. Given $v \in V$ there is $F_v : \Lambda V \to \Lambda V$ defined by $F_v(X) = v \wedge X$. There is also its adjoint $(F_v)^* : \Lambda V \to \Lambda V$ defined by

$$\langle X, F_v Y \rangle = \langle F_v^* X, Y \rangle, \text{ all } X,Y \in \Lambda V.$$

Define $\varphi_v : \Lambda V \to \Lambda V$ by $\varphi_v = F_v + (F_v)^*$.

(2.3) _Let_ V _and_ W _be complex inner product spaces, let_ $v \in V$, $w \in W$ _and consider_ $v + w \in V + W$. _Using the identification_ $\Lambda(V + W) = \Lambda V \otimes \Lambda W$, _we have_

$$\varphi_{v+w}(X \otimes Y) = \varphi_v X \otimes Y + (-1)^k X \otimes \varphi_w Y, \ X \in \Lambda^k V.$$

Proof. The element $v + w$ corresponds to $v \otimes 1 + 1 \otimes w \in \Lambda V \otimes \Lambda W$. Hence

$$F_{v+w}(X \otimes Y) = (v \wedge X) \otimes Y + (-1)^k X \quad (w \wedge Y)$$

$$= F_v(X) \otimes Y + (-1)^k X \otimes F_w(Y),$$

$$F_{v+w} = F_v \otimes 1 + \beta \circ (1 \otimes F_w)$$

where $\beta : \Lambda V \otimes \Lambda W \to \Lambda V \otimes \Lambda W$ maps $X \otimes Y$ into $(-1)^k X \otimes Y$. It may be verified that

$$\begin{aligned}(F_{v+w})^* &= (F_v)^* \otimes 1 + (1 \otimes (F_w)^*) \circ \beta^* \\ &= (F_v)^* \otimes 1 + \beta \circ (1 \otimes (F_w)^*)\end{aligned}$$

since $\beta^* = \beta$. The remark follows.

(2.4) For each $v \in V$ we have $(\varphi_v)^2 = ||v||^2\, I$.

Proof. As an exercise the reader may check this in case $\dim V = 1$. If $\dim V > 1$ split V as the direct sum of orthogonal subspace $V_1 + V_2$ where $\dim V_1 > 0$, $\dim V_2 > 0$ and suppose (2.4) holds for $V_1$ and $V_2$. For $v \in V_1$ and $w \in V_2$ we have

$$\begin{aligned}(\varphi_{v+w})^2(X \otimes Y) &= (\varphi_v)^2 X \otimes Y + (-1)^{k+1} \varphi_v X \otimes \varphi_w(Y) + (-1)^k \varphi_v X \otimes \varphi_w Y \\ &\quad + X \otimes (\varphi_w)^2 Y = (||v||^2 + ||w||^2)(X \otimes Y) \\ &= (||v + w||^2) X \otimes Y.\end{aligned}$$

The remark follows.

Recall that U(n) acts naturally on V.

(2.5) For any $v \in V$ and $g \in U(n)$ we have $\varphi_{gv} \circ g = g \circ \varphi_v$.

Proof. Since $F_v(X) = v \wedge X$ we have

$$g(F_v(X)) = gv \wedge gX = F_{gv}(gX)$$

or $g \circ F_v = F_{gv} \circ g$. Since g is unitary then $g^* = g^{-1}$ and

$$(F_v)^* \circ g^* = g^* \circ (F_{gv})^*, \quad g \circ (F_v)^* = (F_{gv})^* \circ g.$$

Hence $g\varphi_v = \varphi_{gv} g$.

Suppose now that V has an SU-structure given by $\sigma \in \Lambda^n V$; there is the induced operator $\mu : \Lambda V \longrightarrow \Lambda V$.

(2.6) *For each* $v \in V$ *we have* $\varphi_v \mu = \mu \varphi_v$.

Proof. We show first that on $\Lambda^k V$ we have $\tau F_v = (-1)^k (F_v)^* \tau$. Let $X \in \Lambda^k V$. Then

$$\begin{aligned} \langle \tau F_v(X), Y \rangle &= \langle \sigma, v \wedge X \wedge Y \rangle \\ &= (-1)^k \langle \sigma, X \wedge F_v(Y) \rangle = (-1)^k \langle \tau X, F_v(Y) \rangle \\ &= (-1)^k \langle (F_v)^* \tau X, Y \rangle . \end{aligned}$$

Hence $\tau F_v = (-1)^k (F_v)^* \tau$. Then

$$\begin{aligned} (\tau \alpha) F_v &= (-1)^{k+k(k+1)/2} (F_v)^* \tau = (-1)^{k(k-1)/2} (F_v)^* \tau \\ &= (F_v)^* \tau \alpha , \end{aligned}$$

that is, $\mu F_v = (F_v)^* \mu$ and $F_v = \mu^{-1} (F_v)^* \mu$. Then

$$(F_v) \mu = \mu^{-1} (F_v)^* \mu^2 = \mu F_v^*$$

since $\mu^2 = (-1)^{n(n-1)/2} I$. It follows that $\varphi_v \mu = \mu \varphi_v$.

We summarize the situation thus far, combining previous propositions.

(2.7) THEOREM. *Let* $V$ *be a complex inner product space of dimension* $n$ *with given SU-structure* $\sigma \in \Lambda^n V$. *If* $n = 4k + 2$ *then* $\Lambda V$ *is a quaternionic vector space which is* $Z_2$*-graded and for each* $v \neq 0$ *in* $V$ *we have an isomorphism* $\varphi_v : \Lambda^{od} V \approx \Lambda^{ev} V$ *which is quaternionic linear. If* $n = 4k$ *then* $RV$ *is a real vector space which is* $Z_2$*-graded and for each* $v \neq 0$ *in* $V$ *we have a real linear isomorphism* $\varphi_v : R^{od} V \approx R^{ev} V$. *In each case* $\varphi_v$ *commutes with the action of* $SU(n)$.

If $V$ and $W$ are complex inner product spaces, then

$$\begin{aligned} \Lambda^{od}(V + W) &= \Lambda^{ev} V \otimes \Lambda^{od} W + \Lambda^{od} V \otimes \Lambda^{ev} W \\ \Lambda^{ev}(V + W) &= \Lambda^{ev} V \otimes \Lambda^{ev} W + \Lambda^{od} V \otimes \Lambda^{od} W . \end{aligned}$$

Fixing $v \in V$, $w \in W$, and letting $\varphi = \varphi_{v+w}$, $\varphi_1 = \varphi_v$, $\varphi_2 = \varphi_w$, we then have by (2.2):

(2.8) The map $\varphi: \Lambda^{od}(V + W) \longrightarrow \Lambda^{ev}(V + W)$ is given by the matrix

$$\begin{pmatrix} I \otimes \varphi_2 & \varphi_1 \otimes I \\ \varphi_1 \otimes I & -I \otimes \varphi_2 \end{pmatrix}.$$

## 3. Application to bundles.

In this section the constructions of the preceding sections are applied to U(n)-bundles and SU(n)-bundles. For example, given an SU(2k)-bundle $\xi$ there are associated two real vector space bundles $R^{od}(\xi')$ and $R^{ev}(\xi')$ over $D(\xi)$, where $D(\xi)$ is the bundle space of the bundle associated with $\xi$ with fiber the unit ball $D^{8k}$. There is also a linear isomorphism

$$\varphi: R^{od}(\xi')|\partial D(\xi) \approx R^{ev}(\xi')|\partial D(\xi)$$

of the restrictions to the unit sphere bundle $\partial D(\xi)$. Using Atiyah's difference construction, one obtains an element $t(\xi) \in KO(D(\xi), \partial D(\xi))$ where $t(\xi) = d(\Lambda^{ev}(\xi'), \Lambda^{od}(\xi'), \varphi)$. In passing we review the definitions of K-theory and difference classes.

Let $\xi$ be an SU(n)-bundle over a finite CW complex X; we take $\xi$ to be a right principal SU(n)-bundle and denote the bundle space by $E(\xi)$. Fix a complex inner product space V of dimension n with given SU-structure $\sigma \in \Lambda^n V$. Then SU(n) acts on the left on $\Lambda V$ and there is the complex vector space bundle $\Lambda(\xi) \longrightarrow X$, where

$$\Lambda(\xi) = E(\xi) \times \Lambda V/SU(n)$$

and where SU(n) acts on the right on $E(\xi) \times \Lambda V$ by $(e,Y)g = (eg, g^{-1}Y)$. The orbit of (e,Y) under this action is denoted by ((e,Y)). An operator

$\mu$ is defined on $\Lambda(\xi)$ by $\mu((e,Y)) = ((e,\mu Y))$; $\mu$ is well-defined since on V it commutes with the action of SU(n). Replacing $\Lambda V$ by $\Lambda^{od}V'$, $\Lambda^{ev}V'$ respectively in the above, we obtain bundles $\Lambda^{od}(\xi) \longrightarrow X$ and $\Lambda^{ev}(\xi) \longrightarrow X$.

If n = 2 mod 4 then $\mu$ defines a quaternionic bundle structure on $\Lambda(\xi)$, so that in this case we consider $\Lambda(\xi) \longrightarrow X$ a quaternionic vector space bundle. Clearly $\Lambda(\xi)$ splits as the Whitney sum $\Lambda^{ev}(\xi) \oplus \Lambda^{od}(\xi)$.

If n = 0 mod 4, we get a real vector space bundle $R(\xi) \longrightarrow X$, where $R(\xi) = \{x : x \in \Lambda(\xi), \mu x = x\}$. Alternatively,

$$R(\xi) = E(\xi) \times RV/SU(n).$$

Moreover $R(\xi)$ splits as $R^{od}(\xi') \oplus R^{ev}(\xi')$. It also follows from section 1 that as complex bundles $\Lambda(\xi)$ is isomorphic to the complexification of $R(\xi)$; we write this as $\Lambda(\xi) = R(\xi) \otimes_R C$.

We transpose the results of section 2 into bundle notation. Let $\xi$ be an SU(m)-bundle over space X, and $\eta$ an SU(n)-bundle over Y. There is the SU(m) $\times$ SU(n)-bundle

$$\xi \times \eta : E(\xi) \times E(\eta) \longrightarrow X \times Y.$$

By extending the structural group, we also consider $\xi \times \eta$ an SU(m + n)-bundle. We now have from (2.1):

(3.1) <u>There are isomorphisms of vector space bundles</u>

$$\begin{aligned}
R(\xi \times \eta) &= R(\xi) \,\hat{\otimes}_R\, R(\eta), && m = 4k,\ n = 4l \\
\Lambda(\xi \times \eta) &= R(\xi) \,\hat{\otimes}_R\, \Lambda(\eta), && m = 4k,\ n = 4l + 2 \\
\Lambda(\xi \times \eta) &= \Lambda(\xi) \,\hat{\otimes}_R\, R(\eta), && m = 4k + 2,\ n = 4l \\
R(\xi \times \eta) &= \Lambda(\xi) \,\hat{\otimes}_H\, \Lambda(\eta), && m = 4k + 2,\ n = 4l + 2.
\end{aligned}$$

In the above, given vector space bundles $\rho \longrightarrow X$ and $\nu \longrightarrow Y$ we mean by $\rho \hat{\otimes} \nu \longrightarrow X \times Y$ the vector space bundle whose fiber above $(x,y)$ is $\rho^{-1}(x) \otimes \nu^{-1}(y)$. Also in cases 1 and 4, the two bundles are equivalent as real vector space bundles while in cases 2 and 3 they are equivalent as quaternionic vector space bundles.

We next give the significance of the maps $\varphi_v$ of section 2. Let $\xi$ be an SU(n)-bundle over the finite CW complex X. Let

$$D(\xi) = E(\xi) \times D^{2n}/SU(n)$$

where $D^{2n} \subset V$ is the unit disk $\{v : ||v|| \leq 1\}$. Also let

$$\partial D(\xi) = E(\xi) \times S^{2n-1}/SU(n).$$

A point of $D(\xi)$ is an orbit $((e,v))$ where $||v|| \leq 1$ and $((eg,g^{-1}v)) = ((e,v))$. Regard $X = E(\xi)/SU(n)$ as embedded in $D(\xi)$ as the set of $((e,0))$.

There is the SU(n)-bundle $\xi' = f^*(\xi)$ over $D(\xi)$, induced from $\xi$ by the natural map $f : D(\xi) \longrightarrow X$. We then have the complex bundle $\Lambda(\xi') \longrightarrow D(\xi)$; it may be seen that points of $\Lambda(\xi')$ can be taken to be the orbits $((e,v,Y))$ of points of $E(\xi) \times D^{2n} \times \Lambda V$, where

$$((e,v,Y)) = ((eg^{-1}, gv, gY)).$$

Define $\varphi : \Lambda(\xi') \longrightarrow \Lambda(\xi')$ by

$$\varphi((e,v,Y)) = ((e,v, \varphi_v Y)).$$

Note that $\varphi$ is well-defined since

$$\begin{aligned}\varphi((eg^{-1},gv,gY)) &= ((eg^{-1},gv,\ \varphi_{gv}(gY)))\\ &= ((eg^{-1},gv,g\,\varphi_v(Y)))\ \text{by (2.5)}\\ &= \varphi((e,v,Y)).\end{aligned}$$

(3.2) <u>If $\xi$ is an SU(n)-bundle over X, n = 4k + 2, then we have quaternionic vector space bundles $\Lambda^{od}(\xi')$ and $\Lambda^{ev}(\xi')$, and a map $\varphi: \Lambda^{od}(\xi') \longrightarrow \Lambda^{ev}(\xi')$ which above D($\xi$) - X is a quaternionic bundle equivalence. Similarly if n = 4k we get a map $\varphi: R^{od}(\xi') \longrightarrow R^{ev}(\xi')$ which above D($\xi$) - X is a real bundle equivalence.</u>

We can also restate (3.1). For we can identify $D^{2m} \times D^{2n}$ with $D^{2m+2n}$, thus $D(\xi \times \eta)$ with $D(\xi) \times D(\eta)$ and $(\xi \times \eta)'$ with $\xi' \times \eta'$. Then (3.1) becomes

$$R(\xi' \times \eta') \approx R(\xi')\,\hat{\otimes}\,R(\eta'),\ \text{etc.}$$

Also the map $\varphi: R^{od}(\xi' \times \eta') \longrightarrow R^{ev}(\xi' \times \eta')$ can be written in matrix notation exactly as in (2.8).

We now digress to define the groups K(X,A), KO(X,A), KSp(X,A), using a definition that builds in Atiyah's difference construction [7,22]. Fix a pair (X,A) of finite CW complexes; also fix one of the classes of complex, real or quaternionic bundles. Consider triples $(\xi_0, \xi_1, \varphi)$ where $\xi_0$ and $\xi_1$ are vector space bundles over X and $\varphi$ is a vector space isomorphism $\varphi: \xi_1|A \approx \xi_0|A$. Define $(\xi_0, \xi_1, \varphi)$ to be isomorphic to $(\eta_0, \eta_1, \theta)$, written $(\xi_0, \xi_1, \varphi) \approx (\eta_0, \eta_1, \theta)$, if there exist bundle equivalences $\xi_1 \approx \eta_1$ and $\xi_0 \approx \eta_0$ such that commutativity holds in

$$\begin{array}{ccc} \xi_1|A & \xrightarrow{\varphi} & \xi_0|A \\ \downarrow & & \downarrow \\ \eta_1|A & \xrightarrow{\theta} & \eta_0|A. \end{array}$$

Define $(\xi_o, \xi_1, \varphi) \sim (\eta_o, \eta_1, \theta)$ if there exist vector space bundles $\rho$, $\nu$ over X such that

$$(\xi_o \oplus \rho, \xi_1 \oplus \rho, \varphi \oplus 1_\rho) \approx (\eta_o \oplus \nu, \eta_1 \oplus \nu, \theta \oplus 1_\nu).$$

This is checked to be an equivalence relation. Denote by $d(\xi_o, \xi_1, \varphi)$ the equivalence class containing $(\xi_o, \xi_1, \varphi)$ and by K(X,A), KO(X,A), KSp(X,A) the set of equivalence classes.

A unique operation is defined on the set of equivalence classes by

$$d(\xi_o, \xi_1, \varphi) + d(\eta_o, \eta_1, \theta) = d(\xi_o \oplus \eta_o, \xi_1 \oplus \eta_1, \varphi \oplus \theta);$$

a zero element is given by $d(\xi, \xi, 1)$ where $\xi$ is any bundle over X. It is clear that addition is abelian.

It is not difficult to show the existence of negatives, so that the set of equivalence classes becomes an abelian group. For fix $(\xi_o, \xi_1, \varphi)$; given a positive integer n denote by $n_X$ the trivial bundle of dimension n over X. For n large there is an exact sequence of bundles

$$0 \longrightarrow \xi_o \longrightarrow n_X \longrightarrow \rho_o \longrightarrow 0.$$

It may be verified that for n large there exists a linear monomorphism $\xi_1 \longrightarrow n_X$ extending the composition $\xi_1|A \xrightarrow{\varphi} \xi_o|A \longrightarrow n_A$. There is then an exact sequence

$$0 \longrightarrow \xi_1 \longrightarrow n \longrightarrow \rho_1 \longrightarrow 0.$$

Define $\theta : \rho_1|A \longrightarrow \rho_o|A$ so that commutativity holds in

$$\begin{array}{ccccccccc} 0 & \to & \xi_1|A & \to & n_A & \to & \rho_1|A & \to & 0 \\ & & \downarrow \varphi & & \downarrow = & & \downarrow \theta & & \\ 0 & \to & \xi_0|A & \to & n_A & \to & \rho_0|A & \to & 0. \end{array}$$

Then

$$d(\xi_0, \xi_1, \varphi) + d(\rho_0,\rho_1,\theta) = d(n,n,1) = 0.$$

We must compare the above definitions with the usual definitions of $K(X)$, $KO(X)$, $KSp(X)$ in case A is empty [6]. In that case the difference classes can merely be written as $d(\xi_0, \xi_1)$. If we assign to $d(\xi_0, \xi_1)$ the class $\xi_0 - \xi_1$ it is seen that we get an isomorphism of the above group with the classical K-groups.

If X is a finite CW complex with base point $x_0$, then the map $\{x_0\} \to X$ induces $K(X) \to K(\{x_0\})$; it is customary to denote the kernel by $\widetilde{K}(X)$. There is a homomorphism $K(X,x_0) \to \widetilde{K}(X)$ sending $d(\xi_0, \xi_1, \varphi)$ into $\xi_0 - \xi_1$. We assume the fact that this is an isomorphism. If $(X,A)$ is a finite CW pair we also assume a natural isomorphism

$$K(X,A) \xrightarrow{\approx} K(X/A,x_0) \approx \widetilde{K}(X/A);$$

similarly for KO and KSp.

We return now to the main business of this section.

DEFINITION. Let $\xi$ be an $SU(n)$-bundle over a finite CW complex X. Define the Thom space $M(\xi)$ to be $D(\xi)/\partial D(\xi)$. If $n = 4k + 2$, consider the triple $(\Lambda^{ev}(\xi'), \Lambda^{od}(\xi'), \varphi)$ of (3.2), where the bundles are quaternionic bundles over $D(\xi)$ and $\varphi$ is a bundle equivalence over $\partial D(\xi)$. Define

$$s(\xi) = d(\Lambda^{ev}(\xi'), \Lambda^{od}(\xi'), \varphi) \in KSp(D(\xi), \partial D(\xi))$$

or $s(\xi) \in \widetilde{KSp}(M(\xi))$. Similarly if $n = 4k$ we get

$$t(\xi) = d(R^{ev}(\xi'), R^{od}(\xi'), \varphi) \in KO(D(\xi), \partial D(\xi)) = \widetilde{KO}(M(\xi)).$$

Finally given a $U(n)$-bundle $\xi$ over X, one defines

$$\mathcal{L}(\xi) = d(\Lambda^{ev}(\xi'), \Lambda^{od}(\xi'), \varphi) \in K(D(\xi), \partial D(\xi)) = \widetilde{K}(M(\xi))$$

where $\Lambda(\xi')$ and $\varphi$ are considered as complex linear.

Since $\Lambda V$ is the complexification of $RV$ for a vector space V of dimension $4k$ with given SU-structure, we obtain the following.

(3.3) *Let $\xi$ be an $SU(4k)$-bundle. The complexification homomorphism $\widetilde{KO}(M(\xi)) \longrightarrow \widetilde{K}(M(\xi))$ maps to $t(\xi)$ into $\mathcal{L}(\xi)$.*

We now outline very briefly the products in K-theory; for more details, see Atiyah [7] or Solovay [22]. One obtains homomorphisms

$$K(X,A) \otimes K(Y,B) \longrightarrow K(X \times Y, A \times Y \cup X \times B)$$
$$KO(\cdot) \otimes KO(\cdot) \longrightarrow KO(\cdot)$$
$$KO(\cdot) \otimes KSp(\cdot) \longrightarrow KSp(\cdot)$$
$$KSp(.) \otimes KSp(.) \longrightarrow KO(.).$$

Take the first case, and fix $d(\xi_0, \xi_1, \varphi') \in K(X,A)$ and $d(\eta_0, \eta_1, \varphi'') \in K(Y,B)$. It can be shown that $\varphi$ and $\theta$ can be extended to linear homomorphisms $\varphi: \xi_1 \longrightarrow \xi_0$ and $\theta : \eta_1 \longrightarrow \eta_0$. Consider

$$d(\rho_0, \rho_1, \varphi) \in K(X \times Y, A \times Y \cup X \times B)$$

where

$$\rho_1 = \xi_0 \hat{\otimes} \eta_1 + \xi_1 \hat{\otimes} \eta_0, \rho_0 = \xi_0 \hat{\otimes} \eta_0 + \xi_1 \hat{\otimes} \eta_1$$

and where $\varphi: \rho_1 \longrightarrow \rho_0$ is given by the matrix

$$\begin{pmatrix} I \otimes \varphi'' & \varphi' \otimes I \\ \varphi' \otimes I & -I \otimes \varphi'' \end{pmatrix}.$$

In a fashion similar to that of Solovay [22], one sees that

$$d(\rho_0, \rho_1, \varphi) = d(\xi_0, \xi_1, \varphi') \times d(\eta_0, \eta_1, \varphi'')$$

is identified with the usual product of K-theory

$$K(X,A) \otimes K(Y,B) \longrightarrow K(X \times Y,\ A \times Y \cup X \times B),$$
$$\tilde{K}(X/A) \otimes \tilde{K}(Y/B) \longrightarrow \tilde{K}((X/A) \wedge (Y/B)).$$

Let $\xi$ denote a U(m)-bundle over a space X and $\eta$ a U(n)-bundle over Y. Then we identify $M(\xi \times \eta)$ with $M(\xi) \wedge M(\eta)$. The product

$$\tilde{K}(M(\xi)) \otimes \tilde{K}(M(\eta)) \longrightarrow \tilde{K}(M(\xi) \wedge M(\eta)) = \tilde{K}(M(\xi \times \eta))$$

maps $a \otimes b$ into an element denoted by $a \times b$.

(3.4) THEOREM. <u>In</u> $\tilde{K}(M(\xi \times \eta))$ <u>we have</u> $\mathcal{I}(\xi \times \eta) = \mathcal{I}(\xi) \times \mathcal{I}(\eta)$.

This follows from the remarks after (3.2). If $\xi$ is an SU(m)-bundle and $\eta$ an SU(n)-bundle, we have similarly

$$t(\xi \times \eta) = t(\xi) \times t(\eta),\ m = 4k,\ n = 4l$$
$$s(\xi \times \eta) = t(\xi) \times s(\eta),\ m = 4k,\ n = 4l + 2$$
$$t(\xi \times \eta) = s(\xi) \times s(\eta),\ m = 4k + 2,\ n = 4l + 2.$$

4. Thom classes of line bundles.

Suppose that $\xi$ is an SU(2)-bundle over a finite complex X; according to section 3 we receive an element $s(\xi) \in KSp(M(\xi))$. A purpose of this section is to compute $s(\xi)$. Similarly if $\xi$ is a U(1)-bundle over X, we compute $\mathcal{I}(\xi) \in K(M(\xi))$.

So let $\xi$ be an SU(2)-bundle over X. Since SU(2) = Sp(1), then

$\xi$ is an Sp(1)-bundle. Form the join $E(\xi) \circ Sp(1)$, and denote its points by $(1-t)e + th$ where $0 \leq t \leq 1$, $e \in E(\xi)$, $h \in Sp(1)$. Then a principal action of Sp(1) is given by

$$((1-t)e + th)g = (1-t)eg + t \cdot hg.$$

(4.1) <u>The Thom space</u> $M(\xi)$ <u>is canonically isomorphic to</u> $E(\xi) \circ Sp(1)/Sp(1)$.

Proof. Recall that $D(\xi) = E(\xi) \times D^4/Sp(1)$, where $D^4$ is the unit disk in the space H of quaternions. Points of $D(\xi)$ are denoted by $((e,v))$. Define $f' : E(\xi) \times D^4 \longrightarrow E(\xi) \circ Sp(1)$ by

$$f'(e,v) = (1 - |v|)e + |v|(\bar{v}/|v|),$$

and note that f' is well-defined and equivariant with respect to Sp(1)-actions. Passing to orbit spaces, we have a map $f : (D(\xi), \partial D(\xi)) \longrightarrow (E(\xi) \circ Sp(1)/Sp(1), x_0)$ where $x_0$ is the orbit containing all $1 \cdot h$ where $h \in Sp(1)$. We thus get a map $M(\xi) \longrightarrow E(\xi) \circ Sp(1)/Sp(1)$, which we also denote by f. It is checked that f is one-to-one and onto, thus a homeomorphism since all spaces are compact Hausdorff.

(4.2) THEOREM. <u>Let</u> $\xi$ <u>be an</u> SU(2)-<u>bundle over a finite CW complex, and identify</u> $M(\xi)$ <u>with</u> $E(\xi) \circ Sp(1)/Sp(1)$. <u>There is the principal</u> Sp(1)-<u>bundle</u> $E(\xi) \circ Sp(1) \longrightarrow M(\xi)$; <u>denote the associated quaternionic line bundle over</u> $M(\xi)$ <u>by</u> $\eta$. <u>Then</u> $s(\xi) = 1 - \eta$ <u>in</u> $\widetilde{KSp}(M(\xi))$.

Proof. Fix a 2-dimensional complex inner product space V with given SU-structure. Then $\Lambda^{od}V$ and $\Lambda^{ev}V$ can both be identified with the quaternions H. Moreover SU(2) = Sp(1) acts on $\Lambda^{od}V = V = H$ by left multiplication by elements of Sp(1) and SU(2) acts trivially on $\Lambda^{ev}V = H$. Hence by (2.5), $\varphi: \Lambda^{od} \longrightarrow \Lambda^{ev}$ has

$\varphi_{gv}(w) = \varphi_v(g^{-1}\cdot w)$ for $g \in Sp(1)$.

Now $\Lambda^{ev}(\xi')$ is the trivial quaternionic line bundle over $D(\xi)$; let 1 denote the trivial quaternionic line bundle over $E(\xi)\cdot Sp(1)/Sp(1)$. There is the bundle map $F : \Lambda^{ev}(\xi') \longrightarrow 1$ defined by $F(x,w) = (f(x),w)$ for $x \in D(\xi)$ and $w \in H$, where f is defined in the proof of (4.1).

We next obtain a bundle map $G : \Lambda^{od}(\xi') \longrightarrow \eta$. There is

$$G' : (E(\xi) \times D^4) \times H \longrightarrow (E(\xi)\cdot Sp(1)) \times H$$

given by $G'(y,w) = (f'y,w)$ where f' is defined in the proof of (4.1). G' is equivariant with respect to Sp(1)-actions since f' is equivariant. There is induced

$$G : (E(\xi) \times D^4 \times H)/Sp(1) \longrightarrow (E(\xi)\cdot Sp(1)) \times H/Sp(1)$$

or $G : \Lambda^{od}(\xi') \longrightarrow \eta$.

We define finally an isomorphism $\theta : \eta|\{x_0\} \longrightarrow 1|\{x_0\}$, where $x_0$ is the natural base point of $E\cdot Sp(1)/Sp(1)$. The bundle space of $\eta \mid x_0$ consists of all orbits $((h,w))$ for $h \in Sp(1)$ and $w \in H$, where $((h,w)) = ((hg,g^{-1}w))$. Identify the bundle space of $1|\{x_0\}$ with the quaternions H. Define $\theta : \eta|\{x_0\} \longrightarrow 1|\{x_0\}$ by $\theta((h,w)) = \varphi_{h^{-1}}(w)$, where $\varphi_{h^{-1}}$ is the map of section 2; this is well-defined by (2.5). We see that commutativity holds in

$$\begin{array}{ccc} \Lambda^{od}(\xi')|\,\partial D(\xi) & \xrightarrow{G} & \eta|\{x_0\} \\ \downarrow\varphi & & \downarrow\theta \\ \Lambda^{ev}(\xi')|\,\partial D(\xi) & \xrightarrow{F} & 1|\{x_0\}. \end{array}$$

For $\theta G((e,v,w)) = \theta((v^{-1},w)) = \varphi_v(w)$ and

$$F\varphi((e,v,w)) = F((e,v,\varphi_v(w))) = \varphi_v(w),$$

for $||v|| = 1$.

It now follows from general properties of the difference class that

$$f^! : KSp(E \circ Sp(1)/Sp(1), x_0) \longrightarrow KSp(D(\xi), \partial D(\xi))$$

has

$$f^! d(1, \eta, \theta) = d(\Lambda^{ev}(\xi'), \Lambda^{od}(\xi'), \varphi),$$

or identifying the spaces, $1 - \eta = s(\xi)$ in $\widetilde{KSp}(M(\xi))$. The theorem follows.

Consider now the sphere $S^{4n-1}$ as all n-tuples $(\lambda_1, \cdots, \lambda_n)$ of quaternions with $\Sigma|\lambda_i|^2 = 1$. Let Sp(1) act on $S^{4n-1}$ by

$$(\lambda_1, \cdots, \lambda_n)g = (\lambda_1 g, \cdots, \lambda_n g);$$

quaternionic projective space HP(n - 1) is defined to be $S^{4n-1}/Sp(1)$. Thus there is the natural Sp(1)-bundle $\xi_{n-1}$ over HP(n - 1); we also denote by $\xi_{n-1}$ the associated quaternionic line bundle over HP(n - 1). We may regard

$$S^{4n-1} = Sp(1) \circ \cdots \circ Sp(1), HP(n-1) = Sp(1) \bullet \cdots \bullet Sp(1)/Sp(1).$$

We thus have the following corollary.

(4.3) COROLLARY. _The_ Sp(1)-_bundle_ $\xi_{n-1}$ _over_ HP(n - 1) _has_ _Thom_ _space_ HP(n), _and_ $s(\xi_{n-1}) = 1 - \xi_n$ _in_ $\widetilde{KSp}(HP(n))$.

Naturally, entirely similar results hold for complex line bundles.

In particular,

(4.4) <u>The Hopf</u> U(1)-<u>bundle</u> $\rho_{n-1}$ <u>over</u> CP(n - 1) <u>has Thom space</u> CP(n), and $T(\rho_{n-1}) = 1 - \rho_n$ in $\tilde{K}(CP(n))$.

If we apply (4.4) to the trivial complex line bundle $\rho_o$ over a point, then $\mathcal{J}(\rho_o) \in \tilde{K}(S^2)$ is $1 - \rho_1$ where $\rho_1$ is the Hopf bundle over $S^2 = P_1(C)$. In particular $\mathcal{J}(\rho_o)$ is a generator of $\tilde{K}(S^2) \approx Z$.

DEFINITION. Consider the category $\mathcal{C}$ of finite CW complexes with base point, and of base point preserving maps. Denote by $\mathcal{G}$ the category of Z-graded abelian groups and degree preserving homomorphisms. A <u>cohomology theory</u> on $\mathcal{C}$ is a contravariant functor $\mathcal{C} \longrightarrow \mathcal{G}$, assigning to each X a group $\tilde{h}(X) = \sum \tilde{h}^i(X)$ and to each $f : X \longrightarrow Y$ homomorphisms $f^* : \tilde{h}^i(Y) \longrightarrow \tilde{h}^i(X)$, such that

1) if $f,g : X \longrightarrow Y$ are homotopic as base point preserving maps then $f^* = g^*$,

2) given a finite CW pair (X,A), and letting $i : A \subset X$ be inclusion and $\pi : X \longrightarrow X/A$ the natural map, then

$$\tilde{h}^i(X/A) \xrightarrow{\pi^*} \tilde{h}^i(X) \xrightarrow{i^*} \tilde{h}^i(A)$$

is exact,

3) letting SX denote the suspension $S^1 \wedge X$, there exist isomorphisms $\tilde{h}^i(X) \approx \tilde{h}^{i+1}(SX)$ such that if $f : X \longrightarrow Y$ then commutativity holds in

$$\begin{array}{ccc} \tilde{h}^i(Y) & \xrightarrow{\approx} & \tilde{h}^{i+1}(SY) \\ \downarrow f^* & & \downarrow (Sf)^* \\ \tilde{h}^i(X) & \xrightarrow{\approx} & \tilde{h}^{i+1}(SX). \end{array}$$

The cohomology theory is <u>multiplicative</u> if there are homomorphisms

$$\tilde{h}^i(X) \otimes \tilde{h}^j(Y) \longrightarrow \tilde{h}^{i+j}(X \wedge Y)$$

sending $a \otimes b$ into $a \times b$, such that

4) if $a \in \tilde{h}^i(X)$, $b \in \tilde{h}^j(Y)$, $c \in \tilde{h}^k(Y')$, then $(a \times b) \times c = a \times (b \times c)$ in $h^{i+j+k}(X \wedge Y \wedge Y')$,

5) if $T : X \wedge Y \longrightarrow Y \wedge X$ is induced by the map $(x,y) \longrightarrow (y,x)$ of $X \times Y$, and if $a \in h^i(X)$, $b \in \tilde{h}^j(Y)$, then $a \times b = (-1)^{ij} T^*(b \times a)$,

6) there exists an element $\iota \in \tilde{h}^1(S^1)$ such that $\tilde{h}^i(X) \xrightarrow{\approx} \tilde{h}^{i+1}(SX)$ is given by $a \longrightarrow \iota \times a$,

7) given maps $f : X \longrightarrow X'$ and $g : Y \longrightarrow Y'$ and $a \in \tilde{h}^i(X')$, $b \in \tilde{h}^j(Y')$, then $(f \wedge g)^*(a \times b) = f^*a \times g^*b$.

It can be seen that in a multiplicative cohomology theory, the coefficient group $\tilde{h}^*(S^0)$ is a graded associative, anti-commutative ring with unit. The cohomology theory is periodic of period n if $\tilde{h}^i(X) = \tilde{h}^{i+n}(X)$ for all X and i.

It follows from Bott periodicity [9] that corresponding to each generator of $\tilde{K}(S^2)$, we get a periodic cohomology theory $\tilde{K}^*(\cdot)$ of period 2. For define

$$\tilde{K}^{2n}(X) = \tilde{K}(X), \quad \tilde{K}^{2n+1}(X) = \tilde{K}(SX).$$

There is to be defined an isomorphism $K^{2n}(X) \longrightarrow K^{2n+1}(SX)$ or $\tilde{K}(X) \longrightarrow K(S^2X)$. Given a generator $T \in \tilde{K}(S^2)$, periodicity gives such an isomorphism $a \longrightarrow T \times a$ Otherwise put, $\iota \in \tilde{K}^1(S^1) = K(S^2)$ is defined to be T. Hereafter we fix $T \in \tilde{K}(S^2)$ to be $\mathcal{J}(\rho_0) = 1 - \rho_1$.

(4.5) <u>Let</u> $\xi$ <u>denote the</u> SU(4)-<u>bundle over a point. Then</u> $M(\xi) = S^8$ <u>and</u> $t(\xi) \in \widetilde{KO}(S^8)$ <u>is a generator</u>.

Proof. According to (3.3), complexification $\widetilde{KO}(S^8) \longrightarrow \widetilde{K}(S^8)$ maps $t(\xi)$ into $\mathcal{J}(\xi)$. It is sufficient to prove that $\mathcal{J}(\xi)$ is a generator. Regard $\xi$ as $\xi_0 \times \xi_0 \times \xi_0 \times \xi_0$ where $\xi_0$ is the U(1)-bundle over a point. Then

$$S^8 = M(\xi) = M(\xi_0) \wedge M(\xi_0) \wedge M(\xi_0) \wedge M(\xi_0).$$

From (3.4), we have

$$\mathcal{J}(\xi) = \mathcal{J}(\xi_0) \times \mathcal{J}(\xi_0) \times \mathcal{J}(\xi_0) \times \mathcal{J}(\xi_0)$$

in $\widetilde{K}(S^8)$. Since $\mathcal{J}(\xi_0)$ is a generator of $\widetilde{K}(S^2)$, it follows from periodicity that $\mathcal{J}(\xi)$ is a generator of $\widetilde{K}(S^8)$.

There is also the periodic cohomology theory $\widetilde{KO}^*(\cdot)$. Namely define

$$\widetilde{KO}^{8n-i}(X) = \widetilde{KO}(S^i \wedge X)$$

for $i = 0,1,\cdots,7$. The element $i \in \widetilde{KO}^1(S^1) = \widetilde{KO}(S^8)$ is here chosen to be the generator $t(\xi)$ of (4.5).

According to the proof of (4.5), if $\xi$ is the U(n)-bundle over a point, then $\mathcal{J}(\xi)$ is a generator of $\widetilde{K}(S^{2n})$. Also if $\xi$ is an SU(n)-bundle over a point then

1) if $n = 4k$ then $t(\xi)$ is a generator of $\widetilde{KO}(S^{8k})$,

2) if $n = 4k + 2$ then $s(\xi)$ is a generator of $\widetilde{KSp}(S^{8k+4})$.

We assume the following theorem of Dold [13].

(4.6) <u>Suppose that</u> $h^*$ <u>is a multiplicative cohomology theory</u>. <u>Let</u> $\xi$ <u>be an</u> O(n)-<u>bundle over a finite CW complex</u> X. <u>Let</u> $t \in h^n(D(\xi), \partial D(\xi))$ <u>be such that inclusion</u> $i : (D^n_x, \partial D^n_x) \subset (D(\xi), \partial D(\xi))$, <u>where</u> $D^n_x$ <u>is the cell over</u> $x \in X$, <u>has</u> $h^n(D^n_x, \partial D^n_x)$ <u>a free</u> $h^*(pt)$-<u>module with generator</u> $i^*(t)$. <u>Then there</u>

<u>is an isomorphism</u>

$$h^k(X) \approx h^{k+n}(D(\xi), \partial D(\xi))$$

<u>mapping</u> a <u>into</u> $\pi^*a \cdot t$ <u>where</u> $\pi: D(\xi) \to X$.

The reader may supply a proof of (4.6) along the lines of the proof of (7.4).

As a corollary suppose that $\xi$ is an $U(n)$-bundle over X, and let $\mathcal{J}(\xi) \in K(D(\xi), \partial D(\xi))$. Then $i^*\mathcal{J}(\xi)$ is a generator of $\widetilde{K}(D^n_x, \partial D^n_x) = \widetilde{K}(S^{2n})$, and is a generator of the free $K^*(pt)$-module $K^*(S^{2n})$. Hence we get an isomorphism

$$K(X) \approx K(D(\xi), \partial D(\xi)) \approx \widetilde{K}(M(\xi)).$$

By a similar argument, if $\xi$ is an $SU(4k)$-bundle over X, we get an isomorphism

$$KO(X) \approx \widetilde{KO}(M(\xi)),$$

and if $\xi$ is an $SU(4k + 2)$-bundle over X, we get

$$KO(X) \approx \widetilde{KSp}(M(\xi)).$$

5. <u>Cobordism and homomorphisms into K-theory</u>.

In this section we outline the existence of the cobordism theories, and show the existence of natural transformations $\Omega^*_U(.) \to K^*(.)$ and $\Omega^*_{SU}(.) \to KO^*(.)$.

A <u>spectrum</u> M is a sequence

$$M_1, M_2, \cdots, M_{2n}, \cdots$$

of CW complexes with base point, together with base point preserving maps $S^1 \wedge M_n \to M_{n+1}$. Given a finite CW complex X with base point, denote by $[X,M_n]$ the homotopy classes of base point preserving maps $X \longrightarrow M_n$. Given $f : X \longrightarrow M_n$, there is the composition $S^1 \wedge X \xrightarrow{Sf} S^1 \wedge M^n \longrightarrow M_{n+1}$ which we also denote by $Sf : S^1 \wedge X \longrightarrow M_{n+1}$. Define

$$\tilde{H}^n(X;M) = \text{Dir Lim } [S^k \wedge X, M_{n+k}].$$

It is easily checked that $H^n(\cdot;M)$ is a cohomology theory.

Note that it is sufficient to have only $M_2,\cdots,M_{2n},\cdots$ and maps $S^2 \wedge M_{2n} \longrightarrow M_{2n+2}$. For one then defines

$$\tilde{H}^n(X;M) = \text{Dir Lim } [S^{2k-n} \wedge X, M_{2k}].$$

The spectrum M is <u>convergent</u> if each $M_n$ is $(n-1)$-connected. We then have

$$\tilde{H}^n(X;M) \approx [S^k \wedge X, M_{n+k}], \ k \text{ large}.$$

There is a spectrum MSU defined as follows. Let $\eta_n$ denote a universal SU(n)-bundle over a CW complex BSU(n), and let $MSU(n) = M(\eta_n)$. Since $1 + \eta_n$ is an SU(n + 1)-bundle, there is a unique homotopy class of bundle maps $1 + \eta_n \longrightarrow \eta_{n+1}$, also $M(1 + \eta_n) \longrightarrow M(\eta_{n+1})$, hence $S^2 \wedge MSU(n) \longrightarrow MSU(n+1)$. It is also seen that MSU(n) is (2n - 1)-connected. We thus obtain a convergent spectrum MSU, and a cohomology theory

$$\begin{aligned}\tilde{\Omega}^n_{SU}(X) &= \tilde{H}^n(X;MU) \\ &= [S^{2k-n} \wedge X, MSU(k)], \ k \text{ large}\end{aligned}$$

The unique class of bundle maps $\eta_k \times \eta_\ell \longrightarrow \eta_{k+\ell}$ yields a unique homotopy class of maps $MSU(k) \wedge MSU(\ell) \longrightarrow MSU(k+\ell)$, and a product in the cohomology theory $\tilde{\Omega}^*_{SU}(\cdot)$.

It may be seen that $\widetilde{\Omega}^*_{SU}(\cdot)$ is a multiplicative cohomology theory. We do not give details here; it may be helpful for the reader to see [12]. As one part, we define the element

$$\iota \in \Omega^1_{SU}(S^1) = [S^{2k}, MSU(k)]$$

needed for a multiplicative theory. Denote by $\eta$ the SU(k)-bundle over a point, so that $M(\eta) = S^{2k}$. Bundle maps $\eta \to \eta_k$ induce a unique homotpy class of maps

$$S^{2k} = M(\eta) \longrightarrow M(\eta_k) = MSU(k).$$

Let his element represent $\iota \in \Omega^1_{SU}(S^1)$.

Similarly there are multiplicative cohomology theories $\widetilde{\Omega}^*_U(\cdot)$, $\widetilde{\Omega}^*_{Sp}(\cdot)$, $\widetilde{\pi}^*(\cdot)$ given by

$$\Omega^*_U(X) = [S^{2k-n} \wedge X, MU(k)], \text{ k large}$$

$$\widetilde{\Omega}^*_{Sp}(X) = [S^{4k-n} \wedge X, MSp(k), \text{ k large}$$

$$\pi^n(X) = [S^k \wedge X, S^{n+k}], \text{ k large.}$$

Note that in the above constructions we may use for $\eta_k$ a principal SU(k)-bundle over a finite CW complex, the bundle being N-universal for N large.

The natural inclusions

$$1 \subset Sp(k) \subset SU(2k) \subset U(2k)$$

induce maps

$$S^{4k} \longrightarrow MSp(k) \longrightarrow MSU(2k) \longrightarrow MU(2k)$$

and multiplicative transformations

$$\widetilde{\pi}^*(\cdot) \longrightarrow \widetilde{\Omega}^*_{Sp}(\cdot) \longrightarrow \widetilde{\Omega}^*_{SU}(\cdot) \longrightarrow \widetilde{\Omega}^*_{U}(\cdot)$$

of cohomology theories.

There is the element $t_{4k} = t(\eta_{4k}) \in \widetilde{KO}(MSU(4k))$, also $s_{4k+2} = s(\eta_{4k+2}) \in \widetilde{KSp}(MSU(4k + 2))$ as defined in section 3. It follows from section 3 that the map

$$\varphi : MSU(m) \wedge MSU(n) \longrightarrow MSU(m + n)$$

has

$$\varphi^!(t_{4k+4\ell}) = t_{4k} \times t_{4\ell}, \ m = 4k, \ n = 4\ell$$

$$\varphi^!(s_{4k+4\ell+2}) = s_{4k+2} \times t_{4\ell}, \ m = 4k + 2, \ n + 4\ell \text{ etc.}$$

Also the natural map $\theta : S^4 \wedge MSU(4k) \longrightarrow MSU(4k + 2)$ has $\theta^!(s_{4k+2}) = s \times t_{4k}$ where $s \in \widetilde{KSp}(S^4)$ is $s(\eta)$ with $\eta$ the SU(2)-bundle over a point. Similarly $\theta' : S^8 \wedge MSU(4k) \longrightarrow MSU(4k + 4)$ has $(\theta')^!(t_{4k+4}) = t \times t_{4k}$ for appropriate $t \in \widetilde{KO}(S^8)$.

We now define $\mu : \widetilde{\Omega}^n_{SU}(X) \longrightarrow KO^n(X)$. Let $\alpha \in \widetilde{\Omega}^n_{SU}(X)$ be represented by $f : S^{8k-n} \wedge X \longrightarrow MSU(4k)$. Let $\mu(\alpha)$ be the image of $t_{4k}$ in the composition

$$\widetilde{KO}(MSU(4k)) \xrightarrow{f^!} KO(S^{8k-n} \wedge X) \approx \widetilde{KO}^{n-8k}(X) = \widetilde{KO}^n(X).$$

We leave it to the reader to verify the following.

(5.1) THEOREM. The transformation $\mu : \widetilde{\Omega}^*_{SU}(\cdot) \longrightarrow \widetilde{KO}^*(\cdot)$ is a multiplicative transformation of cohomology theories.

We can also define $\mu_s : \widetilde{\Omega}^4_{SU}(X) \longrightarrow \widetilde{KSp}(X)$. Let $f : S^{8k} \wedge X \longrightarrow MSU(4k + 2)$ represent an element $\beta$ of $\widetilde{\Omega}^4_{SU}(X)$. Define $\mu_s(\beta)$ to be the image of $s_{4k+2}$ under the composition

$$\widetilde{KSp}(MSU(4k+2)) \xrightarrow{f^!} KSp(S^{8k} \wedge X) = \widetilde{KSp}(X).$$

(5.2) <u>For each finite CW complex</u> X <u>with base point, commutativity holds in</u>

$$\begin{array}{ccc} & \overset{\mu}{\nearrow} & \widetilde{KO}^4(X) = \widetilde{KO}(S^4 \wedge X) \\ \widetilde{\Omega}^4_{SU}(X) & \mu_s & \approx \uparrow \Phi \\ & \searrow & \widetilde{KSp}(X) \end{array}$$

<u>where</u> $\Phi(\eta) = (1 - \xi_1) \otimes_H \eta$ <u>with</u> $\xi_1$ <u>the Hopf</u> Sp(1)-<u>bundle over</u> $S^4$.

Proof. There is the diagram

$$\begin{array}{ccc} S^4 \wedge (S^{8k-4} \wedge X) & \xrightarrow{S^4 \wedge f} & S^4 \wedge (MSU(4k)) \\ \downarrow & & \downarrow \theta' \\ S^{8k} \times X & \xrightarrow{g} & MSU(4k+2) \end{array}$$

where f and g represent an element of $\Omega^4_{SU}(X)$. Then $(\theta')^!(s_{4k+2}) = s \times t_{4k}$ yields the information necessary to prove the remark.

In an entirely similar fashion there is a multiplicative transformation $\mu_c : \widetilde{\Omega}^*_U(\cdot) \longrightarrow K^*(\cdot)$ sending the element $a \in \widetilde{\Omega}^n_U(X)$ represented by $f : S^{2k-n} \wedge X \longrightarrow MU(k)$ into the image of $\mathcal{J}(\xi_k)$ in

$$\widetilde{K}(MU(k)) \xrightarrow{f^!} \widetilde{K}(S^{2k-n} \wedge X) \approx K^{n-2k}(X) = \widetilde{K}^n(X).$$

There is a natural transformation $\widetilde{\Omega}^*_{SU}(\cdot) \longrightarrow \widetilde{\Omega}^*_U(\cdot)$, and commutativity is seen to hold in

$$\begin{array}{ccc} \widetilde{\Omega}^*_{SU}(\cdot) & \longrightarrow & \widetilde{KO}^*(\cdot) \\ \downarrow & & \downarrow \\ \widetilde{\Omega}^*_U(\cdot) & \longrightarrow & \widetilde{K}^*(\cdot). \end{array}$$

6. The homomorphism $\mu_c$.

After discussing the cohomology of MU(n) and the classical Thom isomorphism theorem, we go on to associate with each element of $H^{**}(BU)$ a homomorphism $\Omega_U^*(X) \longrightarrow H^*(X)$. In terms of these homomorphisms we can characterize the composite

$$\Omega_U^*(X) \xrightarrow{\mu_c} K^*(X) \xrightarrow{ch} H^*(X;Z)$$

where ch is Chern character. In particular, for X a point, the composite

$$\Omega_U^*(pt) \xrightarrow{\mu_c} K^0(pt) = Z$$

is characterized in terms of the classical Todd genus [16] and thus $\mu_c$ is determined on the coefficient groups.

Let $E(\xi)$ be the bundle space of a right principal U(n)-bundle $\xi$ over a space X. The associated sphere bundle is given by

$$E(\xi) \times (U(n)/U(n-1))/U(n) \longrightarrow X.$$

There is an identification

$$E(\xi) \times (U(n)/U(n-1))/U(n) \approx E(\xi)/U(n-1)$$

which identifies the orbit $((e, gU(n-1)))$ on the left with the orbit $((eg))$ on the right, where $e \in E(\xi)$ and $g \in U(n)$. The sphere bundle $S(\xi) \longrightarrow X$ associated with $\xi$ is thus identified with the natural map

$$E(\xi)/U(n-1) \longrightarrow X.$$

If $\xi$ is taken to be a universal bundle for U(n) so that $X = BU(n)$, then $\xi$ is also universal with respect to the subgroup

U(n - 1) of U(n), hence $E(\xi)/U(n-1) = BU(n-1)$. That is, the natural map

$$\rho : BU(n-1) \longrightarrow BU(n)$$

induced by $U(n-1) \subset U(n)$ (see Borel [8]) may be taken to be the sphere bundle over BU(n). Considering the pair $(D(\xi), S(\xi))$, we get the exact cohomology sequence

$$\cdots \longrightarrow H^k(D(\xi), S(\xi)) \longrightarrow H^k(D(\xi)) \longrightarrow H^k(S(\xi)) \longrightarrow \cdots, \text{ or}$$
$$\cdots \longrightarrow \tilde{H}^k(MU(n)) \xrightarrow{i^*} H^k(BU(n)) \xrightarrow{\rho} H^k(BU(n-1)) \longrightarrow \cdots.$$

Using the fact that $\rho$ is an epimorphism with $\rho(c_i) = c_i$, $i < n$, and $\rho(c_n) = 0$, we get the following.

(6.1) *The inclusion* $i : BU(n) \subset MU(n)$ *induces* $i^* : H^*(MU(n)) \longrightarrow H^*(BU(n))$ *which maps* $H^*(MU(n))$ *isomorphically onto the ideal of* $H^*(BU(n))$ *generated by* $c_n$.

Define $v_n \in \tilde{H}^{2n}(MU(n))$ by $i^*(v_n) = c_n$.

Next let $\xi$ be an arbitrary U(n)-bundle over a CW complex X. There is a unique homotopy class of bundle maps $f : E(\xi) \longrightarrow E_{U(n)}$, inducing a unique homotopy class of maps $\tilde{f} : M(\xi) \longrightarrow MU(n)$. Define

$$v(\xi) \in H^{2n}(M(\xi))$$

by $v(\xi) = \tilde{f}^*(v_n)$.

It is easily seen that if $g : E(\zeta) \longrightarrow E(\eta)$ is a bundle map of U(n)-bundles, inducing $\tilde{g} : M(\zeta) \longrightarrow M(\eta)$, then $g^*(v(\eta)) = v(\xi)$.

It may also be seen that if $\xi, \eta$ are U(m), U(n)-bundles over X,Y respectively, then in

$$H^{2m+2n}(M(\xi \times \eta)) \approx H^{2m+2n}(M(\xi) \wedge M(\eta))$$

we have $v(\xi \times \eta) = v(\xi) \times v(\eta)$.

Finally let $\xi$ be the U(1)-bundle over a point, so that $M(\xi) = S^2$. Also we may consider $M(\xi) \subset MU(1) = CP(\infty)$ as the standard embedding $CP(1) \subset CP(\infty)$. Since from (6.1) it follows that $v_1$ is a generator of $H^2(CP(\infty))$, we see that $v(\xi)$ is a generator of $H^2(S^2)$. Using the multiplicative property of the preceding paragraph we see also that if $\xi$ is the U(n)-bundle over a point then $v(\xi)$ is a generator of $H^{2n}(S^{2n})$.

We can now deduce the original theorem of Thom as a corollary of Dold's Theorem (4.6).

(6.2) THOM. _Let_ $\xi$ _be_ _a_ U(n)-_bundle_ _over_ _a_ _finite_ _CW_ _complex_ X. _There_ _is_ _the_ _isomorphism_

$$\varphi : H^k(X) \approx H^{k+2n}(D(\xi), S(\xi))$$

_mapping_ _a_ _into_ $\pi^*(a) \cdot v(\xi)$ _where_ $\pi : D(\xi) \longrightarrow X$.

We also denote $\varphi$ by $\varphi_\xi$ and consider it as an isomorphism $H^k(X) \approx H^{k+2n}(M(\xi))$. Note that if $f : E(\xi) \longrightarrow E(\eta)$ is a U(n)-bundle map, inducing $\tilde{f} : M(\xi) \longrightarrow M(\eta)$ and a map $\bar{f} : X \longrightarrow Y$ of base spaces, then commutativity holds in

$$\begin{array}{ccc} H^k(Y) & \xrightarrow{\bar{f}^*} & H^k(X) \\ \downarrow \varphi & & \searrow \varphi \\ H^{k+2n}(M(\eta)) & \xrightarrow{\tilde{f}^*} & H^{k+2n}(M(\xi)). \end{array}$$

Also suppose that $\xi$ and $\eta$ are U(m),U(n)-bundles respectively. There is

$$\varphi_{\xi \times \eta} : H^*(X \times Y) \longrightarrow H^*(M(\xi \wedge \eta)) \cong H^*(M(\xi) \wedge M(\eta)),$$

and if $a \in H^*(X), b \in H^*(Y)$, then $\varphi_{\xi \times \eta}(a \times b) = \varphi_\xi(a) \times \varphi_\eta(b)$.

There are also Thom isomorphisms $\varphi$ for an arbitrary coefficient ring.

At this stage we assume the ring homomorphism $ch : K^0(X) \longrightarrow H^{ev}(X;Q)$, $Q$ the rationals, for a finite CW complex X. Namely if $\xi$ is a U(n)-bundle over X, from $\sum_{i=1}^{n} \exp t_i$, express as a formal power series in the elementary symmetric functions, and replace by the rational Chern classes $c_1(\xi), \cdots, c_n(\xi)$ respectively. The resulting element of $H^{ev}(X;Q)$ is $ch\, \xi$. By suspension we also get $ch : K^1(X) \longrightarrow H^{odd}(X;Q)$. For the properties of ch, see [6].

(6.3) LEMMA. <u>Let $\xi$ be a U(n)-bundle over a finite CW complex X. Let $\mathcal{J}(\xi) \in \widetilde{K}(M(\xi))$ be as in section 3. Then $\varphi^{-1}(ch\, \mathcal{J}(\xi)) \in H^{ev}(X;Q)$ is the formal power series obtained from</u>

$$\frac{(1 - \exp t_1)\cdots(1 - \exp t_n)}{t_1 \cdots t_n}$$

<u>by replacing the elementary symmetric functions by the Chern classes $c_1(\xi), \cdots, c_n(\xi)$.</u>

Proof. Assign to each $\xi$ the element $r(\xi) = \varphi^{-1}(ch\, \mathcal{J}(\xi)) \in H^*(X;Q)$. Note that if $f : \xi \longrightarrow \eta$ is a bundle map covering $\bar{f} : X \longrightarrow Y$, then $\bar{f}^*(r(\eta)) = r(\xi)$. Also for bundles $\xi$ and $\eta$, we have

$$\mathcal{J}(\xi \times \eta) = \mathcal{J}(\xi) \times \mathcal{J}(\eta), \mathrm{ch}\,(\mathcal{J}(\xi \times \eta)) = \mathrm{ch}\,\mathcal{J}(\xi) \times \mathrm{ch}\,\mathcal{J}(\eta),$$

$$\varphi^{-1}_{\xi \times \eta}(\mathrm{ch}\,\mathcal{J}(\xi \times \eta)) = \varphi^{-1}_{\xi}\ \mathrm{ch}\,\mathcal{J}(\xi) \times \varphi^{-1}_{\eta}\ \mathrm{ch}\,\mathcal{J}(\eta),$$

so that if $\xi$ and $\eta$ are bundles over the same space X, then $r(\xi \oplus \eta) = r(\xi) \cdot r(\eta)$ in $H^*(X;Q)$.

Also let $u(\xi)$ denote the element of $H^*(X;Q)$ obtained from

$$\frac{(1 - \exp t_1)\cdots(1 - \exp t_n)}{t_1 \cdots t_n}$$

by replacing the elementary symmetric functions by $c_1(\xi), \cdots, c_n(\xi)$. It can be seen that $f^*(u(\eta)) = u(\xi)$ and $u(\xi \oplus \eta) = u(\xi) \cdot u(\eta)$. A standard splitting argument shows that $r(\xi) = u(\xi)$ for all $\xi$ if it is true for universal line bundles.

Consider then the Hopf complex line bundle $\rho_n$ over CP(n). We have $M(\rho_n) = CP(n + 1)$ and $\mathcal{J}(\rho_n) = 1 - \rho_{n+1}$ by (4.3). Hence $\mathrm{ch}\,\mathcal{J}(\rho_n) = 1 - \exp t$ where $t = c_1(\rho_{n+1})$, and we must know the Thom class $v(\rho_n) \in H^2(CP(n + 1))$. It follows from (6.1) that $i : CP(n) \subset CP(n + 1)$ has $i^* v(\rho_n) = c_1(\rho_n)$, so that $v(\rho_n) = t$. Hence

$$\varphi^{-1}\ \mathrm{ch}\,\mathcal{J}(\rho_n) = (1 - \exp t)/t.$$

Thus $r(\xi) = u(\xi)$ for the universal line bundles, and it then follows for all $\xi$.

DEFINITION. Consider a partition $\omega = (i_1, \cdots, i_r)$ of positive integers, and let $s = 2i_1 + \cdots + 2i_r$. For each $\alpha \in \Omega^n_U(X,A)$, where (X,A) is a finite CW pair, we define a cohomology class $c_\omega(\alpha) \in H^{s+n}(X,A)$. Namely, let $\alpha$ be represented by

$$f : S^{2k-n} \wedge (X/A) \longrightarrow MU(k),\ k \geq i_j \text{ for all } j$$

Let $c_\omega(\alpha) \in H^{s+n}(X,A)$ denote the image of the product $c_\omega = c_{i_1} \cdots c_{i_k}$ of Chern classes in

$$\begin{array}{ccccc} H^s(BU(k)) & \xrightarrow{\varphi} & \tilde{H}^{s+2k}(MU(k)) & \xrightarrow{f^*} & H^{s+2k}(S^{2k-n} \wedge (X/A)) \\ & & & & \simeq \uparrow S^{2k-n} \\ & & & & H^{s+n}(X,A) = \tilde{H}^{s+n}(X,A), \end{array}$$

that is, $c_\omega(\alpha) = (S^{2k-n})^{-1} f^* \varphi(c_\omega)$. It may be verified that $c_\omega(\alpha)$ is independent of the choice of k.

Similarly given a formal power series $S = \Sigma n_i c_{\omega_i}$ where $n_i$ is rational and $\deg \omega_i \to \infty$ as $i \to \infty$, and given $\alpha \in \Omega_U^n(X,A)$, we get $S(\alpha) \in H^{**}(X,A)$ defined by $S(\alpha) = \Sigma n_i c_{\omega_i}(\alpha)$.

(6.4) THEOREM. <u>For</u> (X,A) <u>a finite CW pair, the composition</u>

$$\Omega_U^{ev}(X,A) \xrightarrow{\mu_c} K^0(X,A) \xrightarrow{ch} H^*(X,A;Q)$$

<u>takes</u> $\alpha$ <u>into</u> $S(\alpha)$ <u>where</u> S <u>is the formal power series in the Chern classes obtained by replacing the elementary symmetric functions in</u>

$$\frac{(1 - \exp t_1)\cdots(1 - \exp t_n)}{t_1 \cdots t_n}$$

<u>by</u> $c_1, \cdots, c_n$ <u>and letting</u> $n \to \infty$.

Proof. It is sufficient to consider a map $f : X/A \to MU(k)$ representing an element $\alpha \in \Omega_U^{2k}(X,A)$. Then

$$\mu_c(\alpha) = f^! \mathcal{J}(\xi_k) \in K(X/A)$$

by section 5, and

$$\begin{aligned}\mathrm{ch}\,\mu_c(\alpha) &= f^*\,\mathrm{ch}\,\mathcal{J}(\xi_k)\\ &= f^*\varphi\varphi^{-1}\,\mathrm{ch}\,\mathcal{J}(\xi_k)\\ &= f^*\,\varphi(S) = S(\alpha).\end{aligned}$$

We now consider

$$\mu_c : \Omega_U^{-2n}(pt) \longrightarrow K^o(pt) = Z.$$

Note that

$$\Omega_U^{-2n}(pt) = \Omega_U^{-2n}(S^o) = [S^{2n+2k}, MU(k)].$$

Suppose now that $M^{2n}$ is a closed differentiable submanifold of $S^{2n+2k}$, with normal bundle $\eta$. Suppose also that $\eta$ has a given reduction of structural group to U(k). The cell bundle N associated with $\eta$ may be identified with the tubular neighborhood of $M^{2n}$. A bundle map $f : \eta \longrightarrow \xi_k$ into the universal U(k)-bundle induces a map $\tilde{f} : M(\eta) \longrightarrow MU(k)$ where $M(\eta) = N/\partial N$. The composition

$$S^{2n+2k} \longrightarrow M(\eta) \xrightarrow{\tilde{f}} MU(k),$$

where the first map shrinks $S^{2n-2k}$ - Int N to a point, represents an element $\alpha$ of $\Omega_U^{-2n}(pt) = [S^{2n+2k}, MU(k)]$. There is the diagram

$$\begin{array}{ccccc} H^{2n+2k}(S^{2n+2k}) & \longleftarrow & H^{2n+2k}(N,\ N) & \xleftarrow{\tilde{f}^*} & H^{2n+2k}(MU(k)) \\ & & \simeq\uparrow\varphi & & \simeq\uparrow\varphi \\ & & H^{2n}(M^{2n}) & \xleftarrow{\tilde{f}^*} & H^{2n}(BU(k)). \end{array}$$

We then see from (6.4) that

$$\mu : \Omega_U^{-2n}(pt) \longrightarrow K^o(pt) = Z$$

maps $\alpha$ into the number $\langle S(\eta), \sigma_{2n} \rangle$, where $\sigma_{2n}$ is the orientation class of $M^{2n}$, and $S(\eta)$ is for $\eta$ the element of $H^*(M^{2n};Q)$ constructed in (6.2).

Now $\Omega_U^{-2n}(pt)$ can be identified with the group $\Omega_{2n}^U$ of closed weakly complex 2n-dimensional manifolds (see [12]). In the above construction one simply puts a suitable complex structure on the stable tangent bundle of $M^{2n}$.

(6.5) COROLLARY. _The composition_

$$\Omega_{2n}^U \approx \Omega_U^{-2n}(pt) \xrightarrow{\mu_c} K^0(pt) = Z$$

_maps a cobordism class_ $[M^{2n}]$ _of closed weakly complex manifolds into the integer_ $(-1)^n \, Td\,[M^{2n}]$, _where_ $Td\,[M^{2n}]$ _is the Todd genus of_ $M^{2n}$ _as in Hirzebruch_.

Proof. Consider a stable tangent bundle $\xi$ for $M^{2n}$, where $M^{2n} \subset S^{2n+2k}$ as above. Since $\xi + \eta$ is trivial, then $\mathcal{J}(\xi + \eta) = \mathcal{J}(\xi)\,\mathcal{J}(\eta) = 1$, hence $\mathcal{J}(\eta) = 1/\mathcal{J}(\xi)$. That is, the image of $[M^{2n}]$ in the integers is $\langle S(\eta), \sigma_{2n}\rangle = \langle S'(\xi), \sigma_{2n}\rangle$, $S'(\xi)$ is generated by

$$P(t_1,\cdots,t_m) = \frac{t_1\cdots t_m}{(1-\exp t_1)\cdots(1-\exp t_m)},$$

m large. We may as well suppose m even. The Todd genus $Td\,[M^{2n}]$ of Hirzebruch is the similar number using

$$Q(t_1,\cdots,t_m) = \frac{t_1\cdots t_m}{(1-\exp(-t_1))\cdots(1-\exp(-t_m))}.$$

Note that $P(t_1,\cdots,t_m) = Q(-t_1,\cdots,-t_m)$. The corollary follows readily.

# CHAPTER II. COBORDISM CHARACTERISTIC CLASSES.

Let $h^*(\cdot)$ be a given multiplicative cohomology theory. The main purpose of section 7 is to give the general sufficient conditions so that we may be able to assign to every Sp(m)-bundle $\xi$ over a finite CW complex X, an element

$$\rho(\xi) = 1 + \rho_1(\xi) + \cdots + \rho_m(\xi)$$

in $h^*(X)$ where $\rho_k(\xi) \in h^{4k}(X)$. Roughly speaking, it is sufficient that we be able to assign suitable classes $\rho_1(\xi)$ for Sp(1)-bundles $\xi$. In order to make such classes, we prove a general theorem of Dold [13].

In section 8, the above generality is applied to the symplectic cobordism theory $\Omega^*_{Sp}(\cdot)$ to assign $\rho_k(\xi) \in \Omega^{4k}_{Sp}(X)$ to every Sp(m)-bundle $\xi$. Since $\rho_1(\xi \oplus \eta) = \rho_1(\xi) + \rho_1(\eta)$, we get

$$\rho_1 : \widetilde{KSp}(X) \longrightarrow \Omega^4_{Sp}(X).$$

For a finite connected CW complex this turns out to embed $\widetilde{KSp}(X)$ additively in $\Omega^4_{Sp}(X)$. Proceeding slightly differently, we define a homomorphism

$$\rho_o : KO(X,A) \longrightarrow \Omega^o_{Sp}(X,A)$$

which embeds KO(X,A) additively as a direct summand of $\Omega^o_{Sp}(X,A)$. There is a similar embedding of KO(X,A) in the special unitary groups $\Omega^o_{SU}(X,A)$.

Quite similarly, there is a homomorphism

$$\underline{c}_o : K(X,A) \longrightarrow \Omega^o_U(X,A)$$

embedding K(X,A) additively as a direct summand.

These are applied in section 10 to determine $K^*(X,A)$ from $\Omega_U^*(X,A)$ and $KO^*(X,A)$ from $\Omega_{Sp}^*(X,A)$. Specifically we have for each n the homomorphism $\mu_c : \Omega_U^{-2n} \longrightarrow K_U^{-2n} = Z$ giving rise to a ring homomorphism $\Omega_U^* \longrightarrow Z$, essentially the classical Todd genus. This allows Z to be considered as a left $\Omega_U^*$ -module. We show that

$$K^*(X,A) \approx \Omega_U^*(X,A) \otimes_{\Omega_U^*} Z.$$

Similarly

$$KO^*(X,A) \approx \Omega_{Sp}^*(X,A) \otimes_{\Omega_{Sp}^*} KO^*(pt).$$

As another application we consider in section 11 the Anderson-Brown-Peterson results concerning the image of $\Omega_*^{fr} \longrightarrow \Omega_*^{SU}$, showing that they can be formally reduced to questions concerning KO-theory solved by J. F. Adams.

7. A theorem of Dold.

In this section we state and prove a theorem of Dold [13] which generalizes to an arbitrary multiplicative cohomology theory the Leray-Hirsch theorem on fiberings. As a consequence we obtain uniqueness and existence theorems for characteristic classes of quaternionic and complex bundles.

Fix once and for all a multiplicative cohomology theory $\tilde{h}(\cdot)$ as in section 4, defined on the category of finite CW complexes with base point. As is well-known, there is generated a multiplicative cohomology theory $h(\cdot)$ on the category of finite CW pairs. For a finite CW pair (X,A), one lets $h(X,A) = \tilde{h}(X/A)$. The external product of section 4 gives rise to an external product

$$h(X,A) \otimes h(Y,B) \longrightarrow h((X \times Y), A \times Y \cup X \times B)$$

sending $a \otimes b$ into $a \times b$. Maps $f : (X,A) \longrightarrow (Y,B)$ give rise to homomorphisms $f^* : h(Y,B) \longrightarrow h(X,A)$. We have that h (point) is $\tilde{h}(S^0)$; hence we call the $h^i$ (pt) the <u>coefficient groups</u>. In terms out in the fashion of Puppe that for each finite CW pair (X,A) there is an exact sequence

$$\cdots \longrightarrow h^n(X,A) \longrightarrow h^n(X) \longrightarrow h^n(A) \longrightarrow h^{n+1}(X,A) \longrightarrow \cdots.$$

Hence $h(\cdot)$ satisfies the Eilenberg-Steenrod axioms except for the dimensional axiom. There is also a cup product

$$h(X,A) \otimes h(X,B) \longrightarrow h(X, A \cup B)$$

sending $a \otimes b$ into $a \cdot b$.

(7.1) <u>Let</u> X <u>be</u> <u>a</u> <u>finite</u> <u>CW</u> <u>complex</u> <u>and</u> <u>let</u> $X^n$ <u>denote</u> <u>its</u> n-<u>skeleton</u>. <u>Define</u> <u>a</u> <u>filtration</u>

$$F^0h(X) \supset F^1h(X) \supset \cdots \supset F^rh(X) \supset \cdots$$

<u>of</u> h(X) <u>by</u> $F^rh(X) = \text{Kernel}\ [i^* : h(X) \longrightarrow h(X^{r-1})]$. <u>Then</u> <u>if</u> $a \in F^rh(X)$ <u>and</u> $b \in F^sh(X)$ <u>we</u> <u>have</u> $a \cdot b \in F^{r+s}h(X)$.

Proof. Consider $X \times X$ as a CW complex using the product of cells. Then

$$(X \times X)^{r+s-1} \subset X^{r-1} \times X \cup X \times X^{s-1}$$

as is easily seen. By exactness we see that $a = j^*(a')$, $b = k^*(b')$ where

$$j^* : h(X,X^{r-1}) \longrightarrow h(X), k^* : h(X,X^{s-1}) \longrightarrow h(X).$$

The element

$$a' \times b' \in h(X \times X, X^{r-1} \times X \cup X \times X^{s-1})$$

then has $\chi^*(a' \times b') = a \times b$ where

$$\chi : X \times X \subset (X \times X, X^{r-1} \times X \cup X \times X^{s-1}).$$

It follows readily from exactness that

$$h(X \times X) \longrightarrow h((X \times X)^{r+s-1})$$

maps $a \times b$ into zero.

Consider the diagonal map $f : X \longrightarrow X \times X$ mapping $x$ into $(x,x)$. There is a cellular map $g : X \longrightarrow X \quad X$ homotopic to $f$, and having $f(X^{r+s-1}) \subset (X \times X)^{r+s-1}$. In

$$\begin{array}{ccc} h(X \times X) & \longrightarrow & h((X \times X)^{r+s-1}) \\ \downarrow g^* & & \downarrow g'^* \\ h(X) & \xrightarrow{m^*} & h(X^{r+s-1}) \end{array}$$

we see that $m^* g^*(a \times b) = m^*(ab) = 0$, hence $ab \in F^{r+s}h(X)$.

(7.2) COROLLARY. _Let_ $X$ _be a finite connected CW complex of dimension_ $n$, _and let_ $x_0 \in X$. _Suppose_ $a_1, \cdots, a_{n+1} \in h(X)$ _are in the kernel of_ $i^* : h(X) \longrightarrow h(x_0)$. _Then_ $a_1 a_2 \cdots a_{n+1} = 0$.

Proof. In the notation of (7.1) we have $a_i \in F^1 h(X)$, hence $a_1 \cdots a_{n+1} \in F^{n+1} h(X) = 0$. Thus $a_1 \cdots a_{n+1} = 0$.

The external product

$$h(X,A) \otimes h(Y) \longrightarrow h(X \times Y, A \times Y)$$

is of particular interest for $Y$ a point. In that case $h(Y)$ is the coefficient group $h(pt)$, which we denote simply by $h$. We also identify $(X \times Y, A \times Y)$ with $(X,A)$ thus obtaining

$$h(X,A) \otimes h \longrightarrow h(X,A).$$

Hence h(X,A) is a right h-module; we denote the image of $a \otimes \omega$ by $a\omega$, where $a \in h(X,A)$ and $\omega \in h$. Similarly we can define $\omega a$ so that h(X,A) is also a left h-module. Associativity of the product implies that in

$$h(X,A) \otimes h(Y,B) \longrightarrow h(X \times Y, A \times Y \cup X \times B)$$

we have $(a\omega) \times b = a \times (\omega b)$ for $a \in h(X,A)$, $\omega \in h$, $b \in h(Y,B)$. We thus obtain a homomorphism

$$h(X,A) \otimes_h h(Y,B) \longrightarrow h(X \times Y, A \times Y \cup X \times B)$$

sending $a \otimes b$ into $a \times b$. The following theorem can be proved just as was a similar theorem in our previous work [10].

(7.3) THEOREM. *Let* $h(\cdot)$ *be a multiplicative cohomology theory. Also let* X *and* Y *be finite CW complexes such that* h(Y) *is a free h-module. Then the homomorphism* $h(X) \otimes_h h(Y) \longrightarrow h(X \times Y)$ *is an isomorphism.*

We can now prove the theorem of Dold [12].

(7.4) DOLD. *Suppose that* $\pi : E \longrightarrow X$ *is a locally trivial fibering with fiber* F, *where* X *and* F *are finite CW complexes. Suppose that* $c_1, \cdots, c_n \in h(E)$ *are such that for each* $x_0 \in X$ *the h-module* $h(\pi^{-1}(x_0))$ *is a free h-module with basis* $i^*(c_1), \cdots, i^*(c_n)$ *where* $i : \pi^{-1}(x_0) \subset E$. *Then* h(E) *is a free* h(X)*-module with basis* $c_1, \cdots, c_n$. *That is, every* $\alpha \in h(E)$ *has a unique representation as*

$$\alpha = \pi^*(a_1) \cdot c_1 + \cdots + \pi^*(a_n) \cdot c_n$$

*for* $a_1, \cdots, a_n \in h(X)$.

Proof. We first prove the theorem in case $\pi$ is trivial, that is $\pi: X \times F \longrightarrow X$ is a projection. It is sufficient to prove this case when X is connected. Fix $x_0 \in X$. There is $i : F \longrightarrow X \times F$ where $i(y) = (x_0, y)$. Denote by $\psi: h(X) \otimes_h h(F) \xrightarrow{\approx} h(X \times F)$ the isomorphism of (7.3), and denote by $\beta: h(X) \otimes_h H(F) \longrightarrow h(F)$ the composition

$$h(X) \otimes_h h(F) \xrightarrow{\psi} h(X \times F) \xrightarrow{i^*} h(F).$$

The inclusion $j : \{x_0\} \subset X$ induces $j^* : h(X) \longrightarrow h(x_0) = h$, and $\beta(a \otimes b) = j^*(a) \cdot b$, where the right hand side uses the structure of h(F) as a left h-module.

Put otherwise, the maps $X \underset{r}{\overset{j}{\leftrightarrows}} \{x_0\}$ induce $h(X) \underset{r^*}{\overset{j^*}{\rightleftarrows}} h$ and a splitting $h(X) = h \oplus \tilde{h}(X)$. If $a = a_1 \oplus a_2$ in this splitting then $\beta(a \otimes b) = a_1 \cdot b$.

Consider the elements $d_1, \cdots, d_n \in h(X) \otimes_h h(F)$ where $d_i = \psi^{-1}(c_i)$, and let $y_i \in h(F)$ be defined by $y_i = \beta(d_i)$. Then $y_1, \cdots, y_n$ is an h-basis for h(F) and

$$d_i = 1 \otimes y_i + \sum x_{jk} \otimes y_k$$

where $x_{jk} \in \tilde{h}(X)$. Let Y denote the column vector whose entries are $1 \otimes y_k$, D the column vector whose entries are $d_i$, I the n by n unit matrix and A the matrix $(x_{ij})$. Then

$$D = (I + A)Y.$$

But it is seen from (7.2) that $A^n = 0$ for n sufficiently large. Hence

$$Y = (I - A + A^2 - A^3 + \cdots)D,$$

and $d_1, \cdots, d_n$ generate $h(X) \otimes_h h(F)$ as an h-module.

Suppose that $B = (b_1, \cdots, b_n)$ is a row vector in $h(X)$ with $BD = 0$. Then $B(I + A) = 0$, hence multiplying by $I - A + A^2 \cdots$ we get $B = 0$. Hence $d_1, \cdots, d_n$ is a basis and the theorem holds in case the fibring is trivial.

Consider next the general case. Let $X'$ be a subcomplex of $X$ and let $E' = \pi^{-1}(X')$. Let $M$ be a free h-module with basis $c'_1, \cdots, c'_n$. Define

$$\tau: h(X') \otimes_h M \longrightarrow h(E')$$

by $\tau(a \otimes c'_i) = \pi^* a \cdot k^* c_i$ where $\pi: E' \longrightarrow X'$ and $k : E' \subset E$. According to the first case, if $\pi: E' \longrightarrow X'$ is trivial then $\tau$ is an isomorphism. We shall show inductively that $\tau$ is an isomorphism for every subcomplex of $X$.

Let $X'$ and $X''$ be subcomplexes of $X$. There is the exact Mayer-Vietoris triangle

$$\begin{array}{ccc} h(X' \cup X'') & \longrightarrow & h(X') + h(X'') \\ & \nwarrow & \\ & & h(X' \cap X''). \end{array}$$

Since $M$ is free, we also have the exact triangle

$$\begin{array}{ccc} h(X' \cup X'') \otimes_h M & \longrightarrow & h(X') \otimes_h M + h(X'') \otimes_h M \\ & \nwarrow & \\ & & h(X' \cap X'') \otimes_h M. \end{array}$$

There is then a commutative diagram

$$\begin{array}{ccccc}
h(X' \cup X'') \otimes_h M & & \longrightarrow & & h(X') \otimes_h M + h(X'') \otimes_h M \\
& \searrow \tau_1 & & \swarrow \tau_2 + \tau_3 & \\
& h(E' \cup E'') & \longrightarrow & h(E') + h(E'') & \\
& \nwarrow & & \swarrow & \\
& & h(E' \cap E'') & & \\
& & \uparrow \tau_4 & & \\
& & h(X' \cap X'') \otimes_h M. & &
\end{array}$$

By the five lemma, if $\tau_2$, $\tau_3$ and $\tau_4$ are isomorphism so is $\tau_1$. Hence if $\tau_2$ is an isomorphism and if $E'' \longrightarrow X''$ is trivial (so that $E' \cap E'' \longrightarrow X' \cap X''$ is also trivial), then $\tau_1$ is an isomorphism. The theorem then follows readily.

We use Dold's theorem as a basic device in constructing characteristic classes. The following two theorems give the generalities.

(7.5) THEOREM. <u>Let</u> $h(\cdot)$ <u>be a multiplicative cohomology theory on the category of finite CW pairs. Suppose for each</u> $n > 0$ <u>there is given an element</u> $\rho_n \in h^4(HP(n))$ <u>such that</u>

(a) $h^*(HP(n))$ <u>is a free h-module with basis</u> $1, \rho_n, (\rho_n)^2, \cdots, (\rho_n)^n$,

(b) <u>inclusion</u> $i : HP(n) \subset HP(n+1)$ <u>has</u> $i^*\rho_{n+1} = \rho_n$.

<u>Then there exists a unique function assigning to each</u> $Sp(m)$<u>-bundle</u> $\xi$ <u>over a finite CW complex</u> $X$ <u>(</u>$m$ <u>arbitrary) an element</u>

$$p(\xi) = 1 + p_1(\xi) + \cdots + p_m(\xi)$$

<u>where</u> $p_k(\xi) \in h^{4k}(X)$, <u>such that</u>

(1) <u>a bundle map</u> $f : \xi \longrightarrow \eta$ <u>covering a map</u> $\bar{f} : X \longrightarrow Y$ <u>of</u>

<u>base spaces has</u> $f^*p(\eta) = p(\xi)$,

(2) <u>if</u> $\xi, \eta$ <u>are</u> Sp(m), Sp(n)-<u>bundles over</u> X <u>respectively, then</u> $p(\xi + \eta) = p(\xi)\cdot p(\eta)$,

(3) <u>if</u> $\xi_n$ <u>is the Hopf</u> Sp(1)-<u>bundle over</u> HP(n) (see section 4), <u>then</u> $p(\xi_n) = 1 + \rho_n$.

Proof. We shall first prove uniqueness. It is clear from (1) and (3) that $p(\xi)$ is uniquely determined for Sp(1)-bundles. Suppose that $\xi$ is an Sp(m)-bundle over X, and that uniqueness holds for Sp(n)-bundles, $n < m$. There is the associated sphere bundle $S(\xi) \longrightarrow X$; moreover Sp(1), the unit sphere of quaternions, acts freely on the right of $S(\xi)$. Let $HP(\xi) = S(\xi)/Sp(1)$, and denote by $\eta$ the Sp(1)-bundle $S(\xi) \longrightarrow HP(\xi)$. Let $\rho \in h^4(HP(\xi))$ be defined by $p = p_1(\eta)$.

There is the natural map $\pi: HP(\xi) \longrightarrow X$, a locally trivial fibering with fiber HP(m - 1). An inclusion $S^{4m-1} \subset S(\xi)$, where $S^{4m-1}$ is a fiber of $S(\xi)$, induces an inclusion $i : HP(m-1) \subset HP(\xi)$. It is then seen that $h^*(HP(m-1))$ is a free h-module with basis 1, $i^*p, \cdots, i^*p^{m-1}$. We may then apply (7.4); in particular $\pi^* : h^*(X) \longrightarrow h^*(HP(\xi))$ is a monomorphism. The bundle $\pi^! \xi$ over $HP(\xi)$ is easily seen to split as $\pi^! \xi = \xi' + \eta$ where $\xi'$ is an Sp(m - 1)-bundle. Hence $p(\pi^! \xi) = \pi^* p(\xi)$ is uniquely determined. Since $\pi^*$ is a monomorphism, $p(\xi)$ is uniquely determined.

We outline without full details the well-known process of showing existence [11]. First of all it follows easily that line bundles have well-defined classes $p_1(\xi)$. Assuming for the moment that existence holds, we would have $\pi^* p(\xi) = (1 + p)\cdot p(\xi')$ as above and $p(\xi') = \pi^* p(\xi)(1 + p)^{-1}$. Since $\xi'$ is an Sp(m - 1)-bundle we would then have $p_m(\xi') = 0$, hence

$$(-1)^m p^m + (-1)^{m-1}\, \pi^* p_1(\xi)\cdot p^{m-1} + \cdots + \pi^* p_m(\xi) = 0.$$

In view of (7.4) we can use the above equation to define $p(\xi)$. Namely define $p_1(\xi), \cdots, p_m(\xi)$ to be the unique elements of $h(X)$ with $\sum_{i=0}^{m} (-1)^i\, \pi^*(p_i(\xi))\cdot p^{m-i} = 0$. It is not difficult to check (1) and (3); we outline the proof of (2).

To show that $p(\xi \oplus \eta) = p(\xi)\cdot p(\eta)$, note first that $HP(\xi) \subset HP(\xi \oplus \eta)$, $HP(\eta) \subset HP(\xi \oplus \eta)$, that $HP(\xi)$ is a deformation retract of $U = HP(\xi \oplus \eta) - HP(\eta)$ and $HP(\eta)$ is a deformation retract of $V = HP(\xi \oplus \eta) - HP(\xi)$. Also $HP(\xi \oplus \eta) = U \cup V$; in this outline we treat U and V as subcomplexes whose union is $HP(\xi)$.

In $h^*(HP(\xi \oplus \eta))$, consider $\sum_{i=0}^{m} (-1)^i\, \pi^*(p_i(\xi))\cdot p^{m-i}$. Upon restriction to U this gives zero. Also consider $\sum_{i=0}^{n} (-1)^j\, \pi^*(p_i(\eta))\cdot p^{n-j}$, which upon restriction to V gives zero. Since $HP(\xi \oplus \eta) = U \cup V$, one sees that

$$\left(\sum_{i=0}^{m} (-1)^i\, \pi^*(p_i(\xi))p^{m-i}\right)\cdot\left(\sum_{j=0}^{n} (-1)^j\, \pi^*(p_j(\eta))p^{n-j}\right) = 0$$

in $h^*(HP(\xi \oplus \eta))$. Also

$$\sum_{i=0}^{m+n} (-1)^k\, \pi^*(p_k(\xi + \eta))\cdot p^{m+n-k} = 0.$$

By Dold's Theorem one sees that

$$p_k(\xi \oplus \eta) = \sum_{i+j=k} p_i(\xi)\cdot p_j(\eta)$$

and $p(\xi \oplus \eta) = p(\xi)\cdot p(\eta)$.

Naturally there is an analogue of (7.5) for unitary bundles.

(7.6) THEOREM. _Let_ $h(\cdot)$ _be a multiplicative cohomology theory on the category of finite CW pairs. Suppose for each_ $n > 0$ _there are_ given elements $\gamma_n \in h^2(CP(n))$ such that

(a) $h^*(CP(n))$ is a free h-module with basis $\gamma_n, (\gamma_n)^2, \ldots, (\gamma_n)^n$,

(b) inclusion $i : CP(n) \subset CP(n+1)$ has $i^*\gamma_{n+1} = \gamma_n$.

Then there exists a unique function assigning to each U(m)-bundle $\xi$ over a finite CW complex X (m arbitrary) an element

$$c(\xi) = 1 + c_1(\xi) + \cdots + c_m(\xi)$$

where $c_k(\xi) \in h^{2k}(X)$, such that

(1) a bundle map $f : \xi \to \eta$ covering a map $\bar{f} : X \to Y$ of base spaces has $\bar{f}^* c(\eta) = c(\xi)$,

(2) if $\xi, \eta$ are U(m),U(n)-bundles over X respectively then $c(\xi \oplus \eta) = c(\xi)\cdot c(\eta)$,

(3) if $\rho_n$ is the Hopf U(1)-bundle over CP(n), then $c(\rho_n) = 1 + \gamma_n$.

Naturally the proof is just as above, based on the fibering $\pi : CP(\xi) \to X$ with fiber $CP(m-1)$.

8. Characteristic classes in cobordism.

In this section we set up central tools for this chapter. Recall that in section 5 we have considered the cohomology theories

$$\Omega^*_{Sp}(\cdot) \to \Omega^*_{SU}(\cdot) \to \Omega^*_U(\cdot)$$

of symplectic, special unitary, unitary cobordism. Given an Sp(m)-bundle $\xi$ over a finite CW complex, we will define characteristic classes $p_k(\xi) \in \Omega^{4k}_{Sp}(X)$; it will sometimes be convenient to use $\Omega^*_{Sp}(\cdot) \to \Omega^*_{SU}(\cdot)$ to consider $p_k(\xi) \in \Omega^{4k}_{SU}(X)$. Also given a U(m)-bundle $\xi$, we will define characteristic classes $c_k(\xi) \in \Omega^{2k}_U(X)$.

We first make some remarks about the above cobordism theories. The best understood of these is $\Omega_U^*(\cdot)$. For example the coefficient ring $\Omega_U^*$ has $\Omega_U^{-2n} \approx \Omega_{2n}^U$, the cobordism group of closed weakly complex manifolds of dimension 2n. Hence $\Omega_U^*$ is a polynomial ring over the integers with one generator in each dimension -2n (Milnor [19], Novikov [21]). In many cases $\Omega_U^*(X)$ can be computed [12]. Turning to $\Omega_{SU}^*(\cdot)$, the additive structure of the coefficient group $\Omega_{SU}^*$ has been determined [12]; few computations have been made for $\Omega_{SU}^*(X)$. The structure of $\Omega_{Sp}^*(\cdot)$ has been hardly touched. The coefficient groups $\Omega_{Sp}^*$ have not been computed; however there is the following partial information (Liulevicious [18]):

| n | n > 0 | 0 | -1 | -2 | -3 | -4 | -5 | -6 |
|---|---|---|---|---|---|---|---|---|
| $\Omega^n_{Sp}$ | 0 | Z | $Z_2$ | $Z_2$ | 0 | Z | $Z_2$ | $Z_2$ |

We now turn to the problem of using (7.5) to give cobordism characteristic classes. We have first to understand MSp(1). Let $\xi_n$ denote the Hopf Sp(1)-bundle over HP(n). According to (4.3), the Thom space $M(\xi_n)$ is identified with HP(n + 1). Letting $n \longrightarrow \infty$, we obtain a universal Sp(1)-bundle $\xi$ over $HP(\infty)$, and

$$MSp(1) = M(\xi) \approx HP(\infty).$$

For each n, the inclusion $i : HP(n) \subset HP(\infty)$ represents an element of [HP(n),MSp(1)]. By suspension we obtain an element

$$p_n \in [S^{4k} \wedge HP(n), MSp(k+1)] = \tilde{\Omega}^4_{Sp}(HP(n)).$$

We also need a Thom homomorphism

$$\mu_Z : \Omega^n_{Sp}(\cdot) \rightarrow H^n(\cdot;Z).$$

As in section 6 we can make the identification

$$MSp(n) \simeq BSp(n)/BSp(n-1),$$

hence we may identify $\tilde{H}^*(MSp(n))$ with the ideal in $H^*(BSp(n))$ generated by the ordinary Chern class $c_{2n} \in H^{4n}(BSp(n))$. Let $\iota_n \in H_{4n}(MSp(n))$ be the element corresponding to $c_{2n}$. Given

$$\alpha \in [S^{4n-k} \wedge (X/A), MSp(n)] = \Omega^k_{Sp}(X,A)$$

represented by a map f, then define $\mu_Z(\alpha) \in H^k(X,A)$ to be the image of $\iota_n$ under the composition

$$\begin{array}{ccc} \tilde{H}^{4n}(MSp(n)) & \xrightarrow{f^*} & H^{4n}(S^{4n-k} \wedge (X,A) \\ & & \simeq \uparrow S^{4n-k} \\ & & \tilde{H}^k(X/A) = H^k(X,A). \end{array}$$

We thus obtain a natural transformation $\mu_Z$ of cohomology theories. It is easily seen that

$$\mu_Z : \Omega^4_{Sp}(HP(n)) \longrightarrow H^4(HP(n))$$

maps $\rho_n$ into a generator of $H^4(HP(n))$.

(8.1) <u>The $\Omega^*_{Sp}$-module $\Omega^*_{Sp}(HP(n))$ is a free $\Omega^*_{Sp}$-module with basis</u> $1, \rho_n, \cdots, (\rho_n)^n$.

Proof. The proof is by induction on n, the proposition being obvious for n = 0. Suppose it is true for n, and consider

$$0 \longrightarrow HP(n) \xrightarrow{i} HP(n+1) \xrightarrow{\pi} S^{4n+4} \longrightarrow 0.$$

There is the exact sequence

$$\cdots \longrightarrow \tilde{\Omega}^*_{Sp}(S^{4n+4}) \longrightarrow \Omega^*_{Sp}(HP(n+1)) \xrightarrow{i^*} \Omega^*_{Sp}(HP(n)) \longrightarrow \cdots.$$

By the induction hypotheses, i* is an epimorphism, hence we get a commutative diagram

$$\begin{array}{ccccccccc} 0 & \longrightarrow & \tilde{\Omega}^*_{Sp}(S^{4n+4}) & \xrightarrow{\pi^*} & \Omega^*_{Sp}(HP(n+1)) & \xrightarrow{i^*} & \Omega^*_{Sp}(HP(n)) & \longrightarrow & 0 \\ & & \downarrow \mu_Z & & \downarrow \mu_Z & & \downarrow \mu_Z & & \\ 0 & \longrightarrow & \tilde{H}^*(S^{4n+4}) & \xrightarrow{\pi^*} & H^*(HP(n+1)) & \xrightarrow{i^*} & H^*(HP(n)) & \longrightarrow & 0. \end{array}$$

Using the notation of (7.1), we have $\rho_{n+1} \in F^4\Omega^*_{Sp}(HP(n+1))$, hence $i^*(\rho_{n+1})^{n+1} = 0$ by (7.1). Hence there exists $\nu \in \tilde{\Omega}^{4n+4}_{Sp}(S^{4n+4})$ with $\pi^*\nu = (\rho_{n+1})^{n+1}$. Then $\mu_Z(\nu)$ is a generator of $H^{4n+4}(S^{4n+4})$. It may be seen by induction on k that $\tilde{\Omega}^k_{Sp}(S^k)$ is a free $\Omega^*_{Sp}$-module with generator any element $\nu$ with $\mu_Z(\nu)$ a generator (see [10, p. 15]). Hence $\tilde{\Omega}^*_{Sp}(S^{4n+4})$ has basis $\nu$. It follows easily from the above diagram that $\Omega^*_{Sp}(HP(n+1))$ is a free module with basis $1, \rho_{n+1}, \cdots, (\rho_{n+1})^{n+1}$.

(8.2) COROLLARY. <u>There exists a unique function assigning to each</u> Sp(m)-<u>bundle</u> $\xi$ <u>over a finite CW complex an element</u>

$$p(\xi) = 1 + p_1(\xi) + \cdots + p_m(\xi)$$

<u>where</u> $p_k(\xi) \in \Omega^{4k}_{Sp}(X)$, <u>such that</u> (1), (2), (3) <u>of</u> (7.5) <u>hold where</u> $\rho_n \in \Omega^4_{Sp}(HP(n))$ <u>is defined as above</u>.

We could equally well define that $p_k(\xi)$ as elements of $\Omega^{4k}_{SU}(X)$. In fact Sp(1) = SU(2) so we could also consider $\rho_n \in \Omega^4_{SU}(HP(n))$. Clearly (8.1) holds for $\Omega^*_{SU}$, so that (7.5) holds in that case. The natural map $\Omega^*_{Sp}(\cdot) \longrightarrow \Omega^*_{SU}(\cdot)$ maps the one $p_k(\xi)$ into the other. In later sections we will consider $p(\xi)$ as in $\Omega^*_{Sp}(X)$ or $\Omega^*_{SU}(X)$, trying to make it clear in each case.

Naturally we may also use (7.6). We have $MU(1) = CP(\infty)$. Thus $i : CP(n) \subset CP(\infty)$ yields an element $\gamma_n \in \Omega^2_U(CP(n))$. Then (7.6) applies to prove the following:

(8.3) COROLLARY. *There exists a unique function which assigns to each* $U(m)$-*bundle* $\xi$ *over a finite CW complex* $X$ *an element*

$$\underline{c}(\xi) = 1 + \underline{c}_1(\xi) + \cdots + \underline{c}_m(\xi)$$

*where* $\underline{c}_k(\xi) \in \Omega^{2k}_U(X)$, *such that* (1), (2) *and* (3) *of* (7.6) *hold, where* $\gamma_n$ *is defined in the above paragraph.*

9. *Characteristic classes in K-theory.*

There was defined in section 5 a homomorphism $\mu : \Omega^*_{SU}(\cdot) \longrightarrow KO^*(\cdot)$; we also denote the composition

$$\Omega^*_{Sp}(\cdot) \longrightarrow \Omega^*_{SU}(\cdot) \xrightarrow{\mu} KO^*(\cdot)$$

by $\mu : \Omega^*_{Sp}(\cdot) \longrightarrow KO^*(\cdot)$. Given an $Sp(m)$-bundle $\xi$ over a finite CW complex $X$, we study $\mu(p_k(\xi)) \in KO^{4k}(X)$. In order to do this, we define characteristic classes $\widetilde{p}_k(\xi) \in KO^{4k}(X)$ and show that $\mu(p_k(\xi)) = \widetilde{p}_k(\xi)$. A similar study is made of the classes $\underline{c}_k(\xi)$ of a $U(m)$-bundle. First we need some generalities about K-theory.

There is a natural ring homomorphism $KO^*(X) \longrightarrow K^*(X)$ which is given on bundles by complexification. There is also

$$ch : K^*(X) \longrightarrow H^*(X;Q)$$

mapping $K^{2k}(X)$ into $H^{ev}(X;Q)$ and $K^{2k+1}(X)$ into $H^{od}(X;Q)$. The natural ring homomorphism given by the composite

$$KO^*(X) \longrightarrow K^*(X) \xrightarrow{ch} H^*(X;Q)$$

is denoted by

$$ph : KO^*(X) \longrightarrow H^*(X;Q).$$

It follows by induction on k that $KO^*(S^k)$ is a free $KO^*$-module with a basis consisting of one element $\vartheta$ of $KO^k(S^k)$, namely any element with ph $\vartheta$ the image of a generator under $H^k(S^k;Z) \longrightarrow H^k(S^k;Q)$.

We also need a little information concerning $KSp(X) = KSp^0(X)$. There is the product

$$KSp(X) \times KSp(X) \longrightarrow KO(X)$$

mapping $(\xi, \eta)$ into the tensor product $\xi \otimes_H \eta$ as in section 3. By neglecting symplectic structure we can regard $\xi$ and $\eta$ as unitary bundles and thus form $\xi \otimes_C \eta$. The <u>complexification homomorphism</u> $KO(X) \longrightarrow K(X)$ <u>maps</u> $\xi \otimes_H \eta$ <u>into</u> $\xi \otimes_C \eta$. <u>This follows from the</u> fact that if V and W are quaternionic vector spaces then

$$(V \otimes_H W) \otimes_R C \approx V \otimes_C W.$$

For instance define a map from the left hand side to the right hand side by

$$(v \otimes_H w) \otimes_R a \longrightarrow (vj \otimes_C wj + v \otimes_C w)a.$$

This may be checked to be well-defined and an epimorphism. A check of dimensions then reveals it to be an isomorphism.

Recall also an isomorphism of Bott [ 9 ],

$$\Phi : \widetilde{KSp}(X) \xrightarrow{\approx} KO^4(X).$$

Namely, $\widetilde{KO}^4(X) = \widetilde{KO}(S^4 \wedge X)$ and given $\eta \in \widetilde{KSp}(X)$ we let $\Phi(\eta) = (1 - \xi_1) \otimes_H \eta$ where $\xi_1$ is the Hopf Sp(1)-bundle over $S^4$. It follows from the above paragraph that ph $\Phi(\eta) = \text{ch}\,\eta$.

(9.1) *The* $KO^*$*-module* $KO^*(HP(n))$ *is a free* $KO^*$*-module with basis* $1, \tilde\rho, \cdots, \tilde\rho^n$ *where* $\tilde\rho$ *is the image of* $1 - \xi_n$ ($\xi_n$ *the Hopf* Sp(1)*-bundle over* HP(n)) *under*

$$\Phi: \widetilde{KSp}(HP(n)) \xrightarrow{\cong} KO^4(HP(n))$$

Proof. We may use the proof of (8.1), making suitable replacements including the replacement of $\mu_Z$ by ph. In order to use that proof, we need ph $\tilde\rho = \text{ch}(2 - \xi_n)$. There is the natural diagram

$$\begin{array}{ccc} & S^{4n+3} & \\ \swarrow & & \searrow \\ CP(2n+1) & \xrightarrow{f} & HP(n) \end{array}$$

from which it follows that $CP(2n+1) = CP(\xi_n)$. Using the properties of $CP(\xi_n)$, it can be seen that $f^! \xi_n = \rho_{2n+1} + \bar\rho_{2n+1}$ in $K(CP(2n+1))$, hence

$$\begin{aligned} f^* \,\text{ch}\, \xi_n &= \text{ch}\, \rho_{2n+1} + \text{ch}\, \bar\rho_{2n+1} \\ &= 2 \cosh t \end{aligned}$$

where t is a generator of $H^2(CP(2n+1))$. Hence

$$\text{ph}\, \tilde\rho = u + \text{terms of higher order}$$

where u is a generator of $H^4(HP(n))$, and ph $\tilde\rho^n = u^n$, a generator of $H^{4n}(HP(n))$. The proof now goes exactly as (8.1).

(9.2) COROLLARY. *There exists a unique function assigning to each* Sp(m)*-bundle* $\xi$ *over a finite CW complex* X *an element*

$$\widetilde{p}(\xi) = 1 + \widetilde{p}_1(\xi) + \cdots + \widetilde{p}_m(\xi)$$

where $\widetilde{p}_k(\xi) \in KO^{4k}(X)$, such that (1) and (2) of (7.5) hold and such that $\widetilde{p}(\xi_n) = 1 + \widetilde{\rho}$.

As promised, we now consider $\mu: \Omega^*_{Sp}(\cdot) \longrightarrow KO^*(\cdot)$, the composition

$$\Omega^*_{Sp}(\cdot) \longrightarrow \Omega^*_{SU}(\cdot) \xrightarrow{\mu} KO^*(\cdot)$$

(9.3) THEOREM. Let $\xi$ denote an $Sp(m)$-bundle over a finite CW complex $X$, and let $p_k(\xi) \in \Omega^{4k}_{Sp}(X)$ and $\widetilde{p}_k(\xi) \in KO^{4k}(X)$ be the classes of (8.2) and (9.2). Then $\mu(p_k(\xi)) = \widetilde{p}_k(\xi)$.

Proof. We have that $\mu(p(\xi))$ and $\widetilde{p}(\xi)$ are elements of $KO^*(X)$ satisfying (1) and (2) of (7.5); this follows since $\mu$ is natural, and also multiplicative. Hence to prove the theorem, it suffices to prove that

$$\mu: \Omega^4_{Sp}(HP(n)) \longrightarrow KO^4(HP(n))$$

maps the $\rho_n$ of (8.1) into the element $\widetilde{\rho}$ of (9.1). Let $\rho'_n$ denote the image of $\rho_n$ in $\Omega^4_{Sp}(HP(n)) \longrightarrow \Omega^4_{SU}(HP(n))$. Then $\rho'_n \in \widetilde{\Omega}^4_{SU}(HP(n))$ and it suffices in view of (5.2) to show that

$$\mu_s : \widetilde{\Omega}^4_{SU}(HP(n)) \longrightarrow \widetilde{KSp}(HP(n))$$

maps $p'_n$ into $1 - \xi_n$. We may consider $BSU(2) = HP(N)$ for $N$ large and that the universal $SU(2)$-bundle $\eta$ is $\xi_N$. Then $s(\eta) \in KSp(MSU(2))$ is given by $MSU(2) = HP(N + 1)$, $s(\eta) = 1 - \xi_{N+1}$ according to (4.2). Now $\rho'_n$ is represented by $i : HP(n) \subset HP(N + 1)$, hence

$$\mu_s(\rho'_n) = i^!(s(\eta)) = 1 - \xi_n.$$

The theorem then follows.

It is convenient to extract from the proof of the preceding theorem an interesting fact. First note that if $\xi$, $\eta$ are Sp(m), Sp(n)-bundles over X respectively then $p_1(\xi \oplus \eta) = p_1(\xi) + p_1(\eta)$. Hence there exists a unique homomorphism

$$p_1 : KSp(X) \longrightarrow \Omega^4_{Sp}(X)$$

taking a bundle $\xi$ into $p_1(\xi)$.

(9.4) THEOREM. <u>The homomorphisms</u>

$$\widetilde{KSp}(X) \xrightarrow{p_1} \widetilde{\Omega}^4_{Sp}(X) \xrightarrow{\mu_s} \widetilde{KSp}(X)$$

<u>have</u> $\mu_s p_1(\eta) = -\eta$ <u>for all</u> $\eta \in \widetilde{KSp}(X)$, <u>where</u> X <u>is</u> <u>a</u> <u>connected</u> <u>finite</u> <u>complex</u>.

Proof. The proof will be made for $\xi - k$ where $\xi$ is an Sp(k)-bundle over X. For the bundle $\xi_n$ over HP(n), that $\mu_s p_1(\xi_n) = 1 - \xi_n$ is just the computation of the proof of (9.3). Hence $\mu_s p_1(\xi_n - 1) = 1 - \xi_n$. We proceed by induction on k. Now let $\xi$ be a Sp(k)-bundle. There is $\pi$: HP( ) $\longrightarrow$ X as in the proof of (7.5) and according to (9.1) and (7.5),

$$\pi^! : KO^4(X) \longrightarrow KO^4(HP(\xi))$$

is a monomorphism, as is then

$$\pi^! : \widetilde{KSp}(X) \longrightarrow \widetilde{KSp}(HP(\xi)).$$

Following the proof of uniqueness of (7.5), we see that $\mu_s p_1(\xi - k) = -(\xi - k)$ by induction on k. Hence $\mu_s p_1(\eta) = -\eta$ for all $\eta$.

We now define a somewhat more functorial form of $p_1$, in particular with no connectedness hypothesis. Consider the diagram

$$\begin{array}{ccccc}
\widetilde{KO}(X) & & \overset{p_o}{\dashrightarrow} & & \Omega^{o}_{Sp}(X) \\
\simeq\big\downarrow S^4 & & & & \simeq\big\downarrow S^4 \\
\widetilde{KO}^4(S^4 \wedge X) & & & & \widetilde{\Omega}^4_{Sp}(S^4 \wedge X) \\
& \underset{\simeq}{\nwarrow}\Phi & & \tilde{p}_1 \nearrow & \\
& & \widetilde{KSp}(S^4 \wedge X) & &
\end{array}$$

and define $p_o : \widetilde{KO}(X) \longrightarrow \Omega^o_{Sp}(X)$, for X a finite complex with base point, by

$$p_o(\eta) = (S^4)^{-1}\, \tilde{p}_1\, \Phi^{-1}\, S^4(\eta).$$

Passing to pairs (X,A), we get

$$p_o : KO(X,A) \longrightarrow \Omega^o_{Sp}(X,A).$$

(9.5) COROLLARY. <u>The homomorphisms</u>

$$KO(X,A) \xrightarrow{p_o} \Omega^o_{Sp}(X,A) \xrightarrow{\mu} KO(X,A)$$

<u>have</u> $\mu\, p_o(\eta) = -\eta$ <u>for every</u> $\eta \in KO(X,A)$, <u>for any finite pair</u> (X,A).

Proof. Consider the diagram

$$\begin{array}{ccccc}
\widetilde{KO}(X) & & \underset{\mu}{\overset{p_o}{\rightleftarrows}} & & \widetilde{\Omega}^{o}_{Sp}(X) \\
\simeq\big\downarrow S^4 & & & & \simeq\big\downarrow S^4 \\
\widetilde{KO}^4(S^4 \wedge X) & & \underset{\mu'}{\longleftarrow} & & \Omega^4_{Sp}(S \wedge X) \\
& \Phi\nwarrow\simeq & & p_1 \nearrow & \\
& & \widetilde{KSP}(S^4 \wedge X) & &
\end{array}$$

We have $\mu p_o = (s^4)^{-1} \mu' p_1 \Phi^{-1} s^4$. It follows from (5.2) and (9.4) that $\mu' p_1 \Phi^{-1} = -\mathrm{id}$, hence $\mu p_o = -\mathrm{id}$.

(9.6) COROLLARY. <u>For every finite CW pair</u> (X,A), KO(X,A) <u>is embedded as a direct summand of</u> $\Omega^o_{Sp}(X,A)$ and also of $\Omega^o_{SU}(X,A)$.

It must be emphasized that no doubt $p_o$ is not multiplicative. Hence we have only embedded KO(X,A) <u>additively</u> in $\Omega^o_{Sp}(X,A)$.

We may now redo the above for unitary bundles. There is a homomorphism

$$\underline{c}_1 : K(X) \longrightarrow \Omega^2_U(X)$$

a periodicity isomorphism $\Phi' : \widetilde{K}(X) \xrightarrow{\approx} K^2(X)$. For a finite connected CW complex the composition

$$\widetilde{K}(X) \xrightarrow{\underline{c}_1} \widetilde{\Omega}^2_U(X) \xrightarrow{\mu_c} K^2(X) \longleftarrow K(X)$$

is the negative of the identity. For any X with base point, define

$$\underline{c}_o : \widetilde{K}(X) \longrightarrow \widetilde{\Omega}^o_U(X)$$

as the composition

$$\begin{array}{ccccc} \widetilde{K}(X) & - - - - - - \overset{\underline{c}_o}{-} - - - \rightarrow & & & \widetilde{\Omega}^o_U(X) \\ \uparrow \cong & & & & \uparrow \cong \\ K^2(S^2 \wedge X) & & & & \Omega^2_U(S^2 \wedge X) \\ & \nwarrow^{\Phi'}_{\cong} & & \nearrow^{\underline{c}_1} & \\ & & K^o(S^2 \wedge X) & & \end{array}$$

There is $\underline{c}_o : K(X,A) \longrightarrow \Omega^o_U(X,A)$ and in

$$K(X,A) \xrightarrow{\underline{c}_o} \Omega^o_U(X,A) \xrightarrow{\mu_c} K(X,A)$$

we have $\mu_c \underline{c}_o = -\mathrm{id}$. Thus K(X,A) is embedded additively in

$\Omega_U^0(X,A)$ as a direct summand.

10. A cobordism interpretation for $K^*(X)$.

In this section we improve upon the results of section 9 by showing how to construct the $Z_2$-graded ring $K^*(X,A)$ knowing only the graded algebra $\Omega_U^*(X,A)$ over the module $\Omega_U^*$. In fact, $K^*(X,A) \simeq \Omega_U^*(X,A) \otimes_{\Omega_U^*} Z$ where Z is a $\Omega_U^*$-ring in a natural way. In a similar fashion, $\Omega_{Sp}^*(X,A)$ determines $KO^*(X,A)$.

There is the homomorphism $\mu_c : \Omega_U^{2n} \longrightarrow K^{2n} = Z$; thus for $\omega \in \Omega_U^{2n}$ we consider $\mu_c(\omega)$ as an integer. Since $\Omega_U^{2n+1} = 0$ (see Milnor [19]), we thus have a ring homomorphism $\mu_c : \Omega_U^* \longrightarrow Z$. Hence we can regard Z as a left $\Omega_U^*$-module by defining $\omega \cdot a$ for $\omega \in \Omega_U^*$ and $a \in Z$ to be the integer $\mu_c(\omega)\cdot a$. For (X,A) a finite pair define

$$\Lambda^*(X,A) = \Omega_U^*(X,A) \otimes_{\Omega_U^*} Z$$

where $\Lambda^*(X,A)$ is regarded as $Z_2$-graded by

$$\Lambda^0(X,A) = \Omega_U^{ev}(X,A) \otimes_{\Omega_U^*} Z, \quad \Lambda^1(X,A) = \Omega_U^{od}(X,A) \otimes_{\Omega_U^*} Z.$$

Alternatively it is easily seen that

$$\Lambda^*(X,A) \approx \Omega_U^*(X,A)/R(X,A)$$

where R(X,A) is the least subgroup of $\Omega_U^*(X,A)$ generated by all $c\cdot\omega - c\cdot\mu_c(\omega)$ for $c \in \Omega_U^*(X,A)$ and $\omega \in \Omega_U^*$.

It is seen that $\Lambda^*(\cdot)$ has many properties of a $Z_2$-graded cohomology theory, in particular all except exactness. It will eventually turn out that $\Lambda^*(\cdot)$ is also exact.

There is the natural epimorphism $\beta : \Omega_U^*(X,A) \longrightarrow \Lambda^*(X,A)$

defined by $\beta(c) = c \otimes 1$. There is seen to be a unique homomorphism

$$\hat{\mu} : \Omega_U^*(X,A) \otimes_{\Omega_U^*} Z \longrightarrow K^*(X,A)$$

of $Z_2$-graded groups with $\hat{\mu}(c \otimes n) = n\mu_c(c)$; existence follows from the fact that $\mu_c$ is multiplicative. Commutativity holds in

$$\begin{array}{ccc} \Omega_U^*(X,A) & \xrightarrow{\beta} & \Lambda^*(X,A) \\ & {\scriptstyle \mu_c}\searrow \quad \swarrow{\scriptstyle \hat{\mu}} & \\ & K^*(X,A) & \end{array} .$$

Here $\Lambda^*(X,A)$ and $K^*(X,A)$ are both considered $Z_2$-graded. There is

$$\hat{c}_o : K^*(X,A) \longrightarrow \Lambda^*(X,A),$$

the composition $K^*(X,A) \xrightarrow{c_o} \Omega_U^*(X,A) \xrightarrow{\beta} \Lambda^*(X,A)$. Moreover the composition

$$K^*(X,A) \xrightarrow{\hat{c}_o} \Lambda^*(X,A) \xrightarrow{\hat{\mu}} K^*(X,A)$$

has $\hat{\mu}\hat{c}_o = -\text{id}$.

(10.1) THEOREM. <u>For every finite CW pair</u> $(X,A)$ <u>we have</u> $\hat{\mu} : \Lambda^*(X,A) \simeq K^*(X,A)$ <u>as</u> $Z_2$<u>-graded rings; hence</u>

$$\Omega_U^*(X,A) \otimes_{\Omega_U^*} Z \simeq K^*(X,A).$$

Proof. We consider first the case in which $H^*(X,A;Z)$ is a free abelian group having only even dimensional elements. In this case it follows from a standard spectral sequence argument [10, p. 49] that

$$\Omega_U^*(X,A) \simeq H^*(X,A) \otimes \Omega_U^*$$

as $\Omega_U^*$-modules. More precisely there exists a homogeneous basis

$\{\alpha_j\}$ for $\Omega_U^*(X,A)$ as an $\Omega_U^*$-module such that $\mu_Z(\alpha_j)$ is a basis for $H^*(X,A)$, where $\mu_Z : \Omega_U^n(X,A) \longrightarrow H^n(X,A)$. It follows from (6.4) that ch $\mu_c(\alpha_j)$ has lead term $\pm\, \mu_Z(\alpha_j)$. It then follows from Atiyah-Hirzebruch [6] that the $\mu_c(\alpha_j)$ generate $K^*(X,A)$ as a free $K^*$-module, where $K^*(X,A)$ is taken as $Z_2$-graded.

We need to compute the kernel of $\mu_c : \Omega_U^*(X,A) \longrightarrow K^*(X,A)$. An element is in this kernel if and only if the coefficients from $\Omega_U^*$ used in expressing this element in terms of the $\alpha_j$ all lie in the kernel of $\mu_c : \Omega_U^* \longrightarrow Z$. Hence Kernel $\mu_c \subset$ Kernel $\beta$, hence $\hat\mu$ is an isomorphism in

$$\begin{array}{ccc} \Omega_U^*(X,A) & \xrightarrow{\ \beta\ } & \Lambda^*(X,A) \\ & {\scriptstyle \mu_c}\searrow \quad \swarrow{\scriptstyle \hat\mu} & \\ & K^*(X,A) & \end{array} .$$

As a second case, consider an $\alpha \in \Lambda^*(X,A)$ with $\hat\mu(\alpha) = 0$ in $K^*(X,A)$ such that there exists a map $f : (X,A) \longrightarrow (Y,B)$ with $H^*(Y,B)$ free abelian with even dimensional generators and with $\alpha = f^*(\beta)$ for some $\beta \in \Lambda^*(Y,B)$; we then show $\alpha = 0$. For consider

$$\begin{array}{ccc} \Lambda^*(Y,B) & \xrightarrow{\ f^*\ } & \Lambda^*(X,A) \\ {\scriptstyle \hat\mu'}\downarrow\uparrow{\scriptstyle \hat c'_o} & & {\scriptstyle \hat\mu}\downarrow\uparrow{\scriptstyle \hat c_o} \\ K^*(Y,B) & \xrightarrow{\ f^!\ } & K^*(X,A). \end{array}$$

Since $\hat\mu_1$ is an isomorphism and $\hat\mu'_1 \hat c'_o = -$ id then $\hat c'_o$ is onto, and $\beta = \hat c'_o(\beta')$ for some $\beta'$. Then $\alpha = \hat c_o(f^! \beta')$, $\hat\mu(\alpha) = -f^! \beta' = 0$.

We see finally that the second case is, roughly speaking, the general case. Let $\gamma \in \Omega_U^{ev}(X,A)$, say $\gamma = \gamma_{2k} + \gamma_{2k+2} + \cdots + \gamma_{2k+2n}$ where $\gamma_{2k+2\ell} \in \Omega_U^{2k+2\ell}(X,A)$. There exists $\beta_{-2} \in \Omega_U^{-2}$ with

$\mu_c(\beta_{-2}) = 1$, say by (6.5). Then

$$\beta(\gamma) = (\gamma_{2k} + \gamma_{2k+2}\cdot\beta_{-2} + \cdots + \gamma_{2k+2\ell}\cdot(\beta_{-2})^{\ell})$$

in $\Lambda^{o}(X,A)$. That is, there exists $\gamma' \in \Omega_U^{2k}(X,A)$ with $\beta(\gamma) = \beta(\gamma')$. Now $\gamma'$ is represented by a map

$$f : S^{2n} \wedge (X/A) \longrightarrow MU(k + n)$$

for n sufficiently large. Then the suspension $S^{2n}(\gamma') \in \tilde{\Omega}^{2k+2n}(S^{2n} \wedge (S/A))$ is in the image of

$$f^* : \tilde{\Omega}_U^*(MU(k + n)) \longrightarrow \tilde{\Omega}_U^*(S^{2n} \wedge (X/A)),$$

hence $S^{2n}\ \beta(\gamma)$ is in the image of

$$f^* : \tilde{\Lambda}^*(MU(k + n)) \longrightarrow \tilde{\Lambda}^*(S^{2n} \wedge (X/A)).$$

If $\hat{\mu} : \tilde{\Lambda}^*(X/A) \longrightarrow K^*(X/A)$ maps $\beta(\gamma)$ into zero, then so does

$$\hat{\mu}: \tilde{\Lambda}^*(S^{2n} \wedge (X/A)) \longrightarrow \tilde{K}^*(S^{2n} \wedge (X/A))$$

map $S^{2n}\ \beta(\gamma)$ into zero, hence by case two we have $S^{2n}\ \beta(\gamma) = 0$ and $\beta(\gamma) = 0$ in $\Lambda^*(X,A)$. That is

$$\hat{\mu}: \Lambda^{o}(X,A) \longrightarrow K^{o}(X,A)$$

is an isomorphism. Similarly $\hat{\mu}: \Lambda^1(X,A) \longrightarrow K^1(X,A)$ is an isomorphism and the theorem follows.

We now point out the changes which must be made in order to relate $\Omega^*_{Sp}(\cdot)$ and $KO^*(\cdot)$. In section 5 we have defined a ring homomorphism

$$\mu : \Omega^*_{Sp}(X,A) \longrightarrow KO^*(X,A),$$

the composition $\Omega^*_{Sp}(\cdot) \longrightarrow \Omega^*_{SU}(\cdot) \xrightarrow{\mu} KO^*(\cdot)$. Specializing to the coefficient group, we get a ring homomorphism

$$\mu : \Omega^*_{Sp} \longrightarrow KO^* = KO^*(pt)$$

In particular we can consider $KO^*$ as a left $\Omega^*_{Sp}$-module, letting

$$\omega \cdot \alpha = \mu(\omega) \cdot \alpha$$

for $\omega \in \Omega^*_{Sp}$ and $\alpha \in KO^*$.

We thus obtain a homomorphism

$$\Omega^*_{Sp}(X,A) \otimes_{\Omega^*_{Sp}} KO^* \xrightarrow{\mu \otimes 1} KO^*(X,A) \otimes_{KO^*} KO^* = KO^*(X,A).$$

(10.2) THEOREM. _For every finite CW pair_ (X,A) _we have_

$$\Omega^*_{Sp}(X,A) \otimes_{\Omega^*_{Sp}} KO^*(pt) \simeq KO^*(X,A).$$

Proof. With a crucial change, the proof is quite similar to the proof of (10.1). The minor changes we leave to the reader, and go directly to the critical point. Note that the proof of (10.1) proceeded in three stages. While the first stage held there is considerable generality for pairs (X,A) with $H^*(X,A)$ free abelian and only having even dimensional elements, it was only necessary in the latter stages to apply it in one particular case only, namely to $(MU(n),\infty)$. There MU(n) could be taken to be the Thom space of a N-universal bundle over an N-classifying space BU(n) for n large. One then needed only in part 1 that

$$\tilde{\Omega}^*_U(MU(n)) \simeq H^*(MU(n)) \otimes \Omega^*_U,$$

or equivalently that

$$\Omega^*_U(BU(n)) \simeq H^*(BU(n)) \otimes \Omega^*_U$$

by (5.3).

A similar fact is all that is used to generalize completely the proof of (10.1) to (10.2). It is convenient to choose a particular model for BSp(n). Namely let $M_{n,N}$ denote all n-dim. quaternionic subspaces of an N-dim. quaternionic space, N large, and take $BSp(n) = M_{n,N}$.

We then need that

$$\Omega^*_{Sp}(BSp(n)) \simeq H^*(BSp) \otimes \Omega^*_{Sp},$$

more precisely that there exists elements $\{\alpha_i\}$ in $\Omega^*_{Sp}(BSp)$ such that $\mu_Z : \Omega^*_{Sp}(BSp(n)) \longrightarrow H^*(BSp)$ has $\{\mu_Z(\alpha_i)\}$ a basis for the free abelian group $H^*(BSp(n))$. It then will follow, using the methods of [10, p. 49], that $\Omega^*_{Sp}(BSp(n) \simeq H^*(BSp) \otimes \Omega^*_{Sp}$.

The universal bundle over BSp(n) has Chern classes $c_2, c_4, \cdots, c_{2n}$. It is known that $H^*(BSp(n)) = H^*(M_{n,N})$ has a basis consisting of polynomials $\beta_i$ in the Chern classes. We shall see that every $\beta_i$ is in the image of

$$\Omega^*_{Sp}(BSp(n)) \longrightarrow H^*(BSp(n)).$$

It is sufficient to show that every $c_{2k}$ is in this image.

Consider the natural transformation

$$\mu_Z : \Omega^*_{Sp}(\cdot) \to H^*(\cdot).$$

It may be verified that if $\xi$ is a Sp(m)-bundle over X then $\mu_Z$ maps $p_k(\xi) \in \Omega^{4k}_{Sp}(X)$ into $\pm c_{2k} \in H^{4k}(X)$. In particular taking $\xi$ to be universal bundle over BSp(n) we get that

$$\Omega^*_{Sp}(BSp(n)) \longrightarrow H^*(BSp(n))$$

is an epimorphism. The theorem then follows:

11. Mappings into spheres.

In preceding sections, we have indicated connections between $KO^*(\cdot)$ and $\Omega^*_{Sp}(\cdot)$, also between $KO^*(\cdot)$ and $\Omega^*_{SU}(\cdot)$. The latter is the more fruitful because $\Omega^*_{SU}(\cdot)$ is better understood than $\Omega^*_{Sp}(\cdot)$; in this section we prove a theorem which illustrates this point. The theorem arises from an attempt to understand a theorem of D. Anderson-Brown-Peterson [4] from other points of view. As we consider the question here, which homotopy classes of maps $f : S^{8n+k} \longrightarrow S^{8n}$ induce a non-trivial $f^* : \tilde{\Omega}^*_{SU}(S^{8n}) \longrightarrow \tilde{\Omega}^*_{SU}(S^{8n+k}))$ It follows from our formalism, and some information on $\Omega^*_{SU}$, that $f^*$ is non-trivial if and only if $f^! : KO(S^{8n}) \longrightarrow KO(S^{8n+k})$ is non-trivial. Then one uses the results of Adams [3] concerning when $f^!$ is non-trivial.

We need some information concering the coefficient ring $\Omega^*_{SU}$. Recall that $\Omega^*_{SU}(S^{8n})$ is a free $\Omega^*_{SU}$-module with a generator $\nu_{8n}$; also $\mu : \Omega^*_{SU}(S^{8n}) \longrightarrow KO^*(S^{8n})$ is an epimorphism by (9.5). Hence $\mu(\nu_{8n}) \in KO^{8n}(S^{8n}) = KO^o(S^{8n}) = Z$ must be a generator, i.e. $\mu(\nu_{8n}) = \pm 1$. Consider

$$\widetilde{KO}(S^{8n}) \xrightarrow{p_o} \tilde{\Omega}^o_{SU}(S^{8n}) \xrightarrow{\mu} \widetilde{KO}(S^{8n}) = Z,$$

where $\mu p_o = -\mathrm{id}$. If $\alpha$ is a generator of $\widetilde{KO}(S^{8n})$, then $\mu p_o(\alpha) = \pm 1$. However $p_o(\alpha) = \beta_{-8n} \cdot \nu_{8n}$ where $\beta_{-8n} \in \Omega^{-8n}_{SU}$ and $\mu p_o(\alpha) = \mu(\beta_{-8n}) \cdot \mu(\nu_{8n})$. Considering $\mu : \Omega^{-8n}_{SU} \longrightarrow KO^{-8n}(pt) = Z$ as having integer values, we get $\mu(\beta_{-8n}) = \pm 1$. We need now the following lemma.

(11.1) LEMMA. Let $\beta_{-8n} \in \Omega^{-8n}_{SU}$ have $\mu(\beta_{-8n}) = \pm 1$. Then

$\beta_{-8n}$ <u>is not a divisor of zero in the ring</u> $\Omega^*_{SU}$.

Proof. The proof is based on our previous paper [12]. First we have to convert the statement to one in terms of bordism. According to section 5,

$$\begin{array}{ccc} \Omega^*_{SU}(\cdot) & \xrightarrow{\mu} & KO^*(\cdot) \\ \downarrow & \mu_c & \downarrow \\ \Omega^*_U(\cdot) & \longrightarrow & K^*(\cdot) \end{array}$$

commutes, hence $\beta'_{-8n} \in \Omega^{-8n}_U$ has $\mu_c(\beta'_{8n}) = \pm 1$. Using the isomorphism $\Omega^U_{8n} \approx \Omega^{-8n}_U$, the element $[M^{8n}] \in \Omega^U_{8n}$ corresponding to $\beta'_{8n}$ has Todd genus $T[M^{8n}] = \pm 1$ according to (6.5).

It is then sufficient to switch to bordism. Denote by $\Omega^{SU}_*$ the bordism ring of closed SU-manifolds (denoted by $\Gamma_*$ in [12]). We must prove that <u>if</u> $[M^{8n}] \in \Omega^{SU}_{8n}$ <u>has Todd genus</u> $T[M^{8n}] = \pm 1$, <u>then</u> $[M^{8n}]$ <u>is not a zero divisor in</u> $\Omega^{SU}_*$. In order to prove this we recall some facts [12]. There is a boundary operator $\partial: \Omega^U_{2n} \longrightarrow \Omega^U_{2n-2}$ taking $[W^{2n}]$ into $[V^{2n-2}]$ where $V^{2n-2} \subset W^{2n}$ is dual to the ordinary Chern class $c_1(V^{2n-2})$. Moreover

$$\mathrm{Im}[\Omega^{SU}_* \longrightarrow \Omega^U_*] \supset \mathrm{Im}[\partial : \Omega^U_* \to \Omega^U_*],$$

and $\mathrm{Im}\,\Omega^{SU}_*/\mathrm{Im}\,\partial$ is a polynomial algebra over $Z_2$ with generators in each dimension 8k. Also all torsion of $\Omega^{SU}_*$ is of the form $[W^{8m}][\overline{S}^1]$ or $[W^{8m}][\overline{S}^1][\overline{S}^1]$ where $[W^{8m}] \in \Omega^{SU}_*$ represents a non-zero element of $\mathrm{Im}\,\Omega^{SU}_*/\mathrm{Im}\,\partial$. Finally if $[M^{8n}] \in \Omega^{SU}_{8n}$ has odd Todd genus, then $[M^{8n}]$ represents a non-zero element of $\mathrm{Im}\,\Omega^{SU}_*/\mathrm{Im}\,\partial$ [12, p. 70].

We can now prove our assertion. Suppose $[M^{8n}]$ has odd Todd genus and that $[W^k] \in \Omega^{SU}_k$ has $[M^{8n}][W^k] = 0$. Then $[W^k]$ is a torsion element, for $\Omega^{SU}_* \longrightarrow \Omega^U_*$ has only torsion in its kernel and $\Omega^U_*$ is a polynomial algebra. Hence we may suppose $[W^k] = [V^{8m}][\bar{S}^1]$ or $[W^k] = [V^{8m}][\bar{S}^1][\bar{S}^1]$. Then $[M^{8n}][V^{8m}]$ represents 0 in $\mathrm{Im}\,\Omega_*^{SU}/\mathrm{Im}\,\partial$. Since this is a polynomial algebra, then $[V^{8m}]$ represents zero in $\mathrm{Im}\,\Omega_*^{SU}/\mathrm{Im}\,\partial$ and $[W^k] = 0$. The lemma follows.

(11.2) THEOREM. <u>Suppose that</u> X <u>is a finite CW complex with</u> $\widetilde{\Omega}^*_{SU}(X)$ <u>a free</u> $\Omega^*_{SU}$<u>-module. If</u> $f : X \longrightarrow S^{8n}$ <u>then</u> $f^* : \widetilde{\Omega}^*_{SU}(S^{8n}) \longrightarrow \widetilde{\Omega}^*_{SU}(X)$ <u>is non-trivial if and only if</u> $f^! : \widetilde{KO}(S^{8n}) \longrightarrow \widetilde{KO}(X)$ <u>is non-trivial.</u>

Proof. Consider the diagram

$$\begin{array}{ccc}
\widetilde{\Omega}^{8n}_{SU}(S^{8n}) & \xrightarrow{f^*} & \widetilde{\Omega}^{8n}_{SU}(X) \\
\downarrow \mu & & \downarrow \mu \\
\widetilde{KO}(S^{8n}) & \xrightarrow{f^!} & \widetilde{KO}(X) \\
\downarrow p_o & & \downarrow p_o \\
\widetilde{\Omega}^o_{SU}(S^{8n}) & \xrightarrow{f^*} & \widetilde{\Omega}^o_{SU}(X).
\end{array}$$

Suppose that $f^* : \widetilde{\Omega}^*_{SU}(S^{8n}) \longrightarrow \widetilde{\Omega}^*_{SU}(X)$ is non-trivial. Now $\widetilde{\Omega}^*_{SU}(S^{8n})$ is a free $\Omega^*_{SU}$-module with basis $\nu_{8n} \in \widetilde{\Omega}^{8n}_{SU}(S^{8n})$. According to the first of this section, $p_o\,\mu(\nu_{8n}) = \beta_{-8n}\cdot\nu_{8n}$ where $\beta_{-8n} \in \Omega^{-8n}_{SU}$ has $\mu(\beta_{-8n}) = \pm 1$. Then

$$f^* p_o\,\mu(\nu_{8n}) = \beta_{-8n}\cdot f^*(\nu_{8n}).$$

Since $f^*(\nu_{8n})$ is a non-zero element of the free $\Omega^*_{SU}$-module $\widetilde{\Omega}^*_{SU}(X)$, it follows from (11.1) that $\beta_{-8n}\cdot f^*(\nu_{8n}) \neq 0$. Hence $f^! \neq 0$ and the theorem follows.

Note that in particular the theorem holds for $X = S^{8n+k}$.

(11.3) COROLLARY. <u>Suppose that</u> X <u>is a finite CW complex with base point such that</u> $\widetilde{\Omega}^*_{SU}(X)$ <u>is a free</u> $\Omega^*_{SU}$<u>-module. In the diagram</u>

$$\begin{array}{ccc} \widetilde{\pi}^j(X) & \xrightarrow{\hat{s}} & \widetilde{KO}^j(X) \\ & s \searrow \quad \nearrow \mu & \\ & \widetilde{\Omega}^j_{SU}(X) & \end{array}$$

we have Image s $\approx$ Image s.

Proof. It is sufficient to show that Kernel $\hat{s}$ = Kernel s. If $\alpha \varepsilon \widetilde{\pi}^j(X)$ is represented by $f : S^{8n-j} \wedge X \longrightarrow S^{8n}$ then by (11.2) $f^*(\nu_{8n}) = 0$ if and only if $f^! \mu(\nu_{8n}) = 0$. The corollary follows.

(11.4) ANDERSON-BROWN-PETERSON. The image $\pi^{-j}(pt) \longrightarrow \Omega^{-j}_{SU}(pt)$ is $Z_2$ if $j = 8m + 1$ or $8m + 2$, 0 otherwise.

Proof. Apply (11.3) to $X = S^0$. Then $Im\ [\pi^{-j}(pt) \longrightarrow \Omega^{-j}_{SU}(pt)] = IM\ [\pi^{-j}(pt) \longrightarrow KO^{-j}(pt)]$. According to a result of Adams [3], the right hand side is as stated and the assertion follows. In a later section we give a bordism proof of the theorem of Adams.

# CHAPTER III. U-MANIFOLDS WITH FRAMED BOUNDARIES

In this chapter we shift from our very general point of view of the previous chapters to some very concrete problems on the relationship between U-bordism and K-theory. In section 12 we consider the bordism group $\Omega^U_n$ of closed U-manifolds of dimension n; the elements of $\Omega^U_n$ are the bordism classes $[M^n]$ of closed differentiable manifolds with a given complex structure on the stable tangent bundle. In section 13 we begin to study the numbers

$$x[M] = \langle ch\, x \cdot T^{-1}(M), \sigma(M) \rangle$$

where $x \in K(M)$, $T^{-1}(M)$ is the Todd polynomial of M and $\sigma(M) \in H_*(M)$ is the orientation class. In particular there are the integers $\underline{s}_\omega[M]$ where the $\underline{s}_\omega \in K(M)$ are certain K-theory characteristic classes of the stable tangent bundle. In section 14 we give the proof of Stong [23] that every homomorphism $\Omega^U_{2n} \longrightarrow Z$ is an integral linear combination of the $\underline{s}_\omega$; this theorem has also been proved by Hattori [15].

In section 15 we shift to the compact U-manifolds M with stably framed boundary; we call such a manifold a (U,fr)-manifold. Such a manifold M has a complex structure on its stable tangent bundle $\tau$ together with a compatible framing of the restriction $\tau | \partial M$ to the boundary. Such manifolds have Chern classes and Chern numbers, hence also a Todd genus Td [M] which is now a rational number. It is proved that if $M^{2n}$ is a compact (U,fr)-manifold, then there exists a closed U-manifold with the same Chern numbers if and only if Td $[M^{2n}]$ is an integer.

In section 16 we consider bordism classes of compact (U,fr)-manifolds $M^n$; these may be identified with elements of the homotopy group

$\pi_{n+2k}(MU(k)/S^{2k})$, k large. For $n > 0$ there is the short exact sequence

$$0 \longrightarrow \Omega^{U}_{2n} \longrightarrow \Omega^{U,fr}_{2n} \longrightarrow \Omega^{fr}_{2n-1} \longrightarrow 0.$$

The homomorphism $Td : \Omega^{U,fr}_{2n} \longrightarrow Q$ then gives rise to a homomorphism

$$E : \Omega^{fr}_{2n-1} \longrightarrow Q/Z,$$

which turns out to be equal to a well-known homomorphism

$$e_c : \Omega^{fr}_{2n-1} \longrightarrow Q/Z$$

of J. F. Adams [1]. Thus we obtain a complete description of the image of

$$Td : \Omega^{U,fr}_{2n} \longrightarrow Q,$$

and considerable knowledge concerning the bordism groups $\Omega^{U,fr}_{2n}$ and $\Omega^{SU,fr}_{2n}$.

12. The U-bordism groups $\Omega^{U}_{*}$.

Let $M^n$ denote a differentiable manifold and let $\tau$ denote its tangent bundle. We shall call the Whitney sum $\tau + (2k - n)$ of $\tau$ and the trivial $(2k - n)$-bundle the stable tangent bundle of $M^n$, where $2k - n \geq 2$. Note that $\tau + (2k - n)$ is a real 2k-bundle with space

$$E(\tau + (2k - n)) = E(\tau) \times R^{2k-n}.$$

A U-structure $\Phi$ on $M^n$ is a homotopy class of maps

$$J : E(\tau + (2k - n)) \longrightarrow E(\tau + (2k - n))$$

each of which maps each fiber linearly onto itself and has $J^2 = -$ identity (see [12]).

Given such an operator J on $E(\tau) \times R^{2k-n}$, there is induced an operator J' on $E(\tau) \times R^{2k-n} \times R^2$ given by $J' = J \times J_o$ where $J_o : R^2 \longrightarrow R^2$ is given by $J_o(s,t) = (-t,s)$. It may be seen from this that the giving of a U-structure $\Phi$ is independent of the precise value of k at least if $2k - n \geq 2$ (see [12, p. 16]). Similarly if $E(\tau)$ admits such a natural operator J, then $M^n$ receives a natural U-structure. Thus every almost complex manifold and hence every complex analytic manifold also has a natural U-structure.

A U-manifold $(M^n, \Phi)$ is a pair consisting of a differentiable manifold $M^n$ and a U-structure $\Phi$ on $M^n$. Often we take the U-structure for granted and denote the U-manifold simply by $M^n$.

Fix a U-manifold $M^n$ and let J be a representative of the U-structure $\Phi$. Then J converts $\tau + (2k - n)$ into a complex vector space bundle. This complex vector space bundle has Chern classes, which are denoted by

$$c_k(M^n) \in H^{2k}(M^n), \; k = 0,1,2,\cdots.$$

Moreover $\tau + (2k - n)$ receives a natural orientation as a complex bundle. Now if we take two different representatives of $\Phi$ it is not hard to see that we obtain the same Chern classes $c_k(M^n)$ and the same orientation for $\tau + (2k - n)$. Since $R^{2k-n}$ has a preferred orientation, the tangent bundle $\tau$ thus receives a natural orientation. If $M^n$ is compact, denote by

$$\sigma(M^n) \in H_n(M^n, \partial M^n)$$

the orientation class.

Every closed U-manifold $M^{2n}$ has Chern numbers. Namely given positive integers $i_1, i_2, \cdots, i_p$ with $i_1 + \cdots + i_p = n$ there is the integer

$$c_{i_1}c_{i_2}\cdots c_{i_p}[M^{2n}] = \langle c_{i_1}(M^{2n})\cdots c_{i_p}(M^{2n}), \sigma(M^{2n}) \rangle,$$

the value of the cup product $c_{i_1}\cdots c_{i_p} \in H^{2n}(M^{2n})$ on the orientation class $\sigma \in H_{2n}(M^{2n})$.

We shall be first of all concerned in this chapter with the problem of Milnor [24] and Hirzebruch.

PROBLEM. Suppose that n is given and that for each partition $\{i_1,\dots,i_p\}$ of n we are given an integer $a_{i_1,\dots,i_p}$. What are necessary and sufficient conditions that there exists a closed U-manifold $M^{2n}$ with

$$c_{i_1}\cdots c_{i_p}[M^{2n}] = a_{i_1,\dots,i_p}$$

for each $\{i_1,\dots,i_p\}$?

It is convenient to have at hand the U-bordism groups $\Omega_n^U$. We shall not give complete definitions (see [12]) but simply a quick sketch. It is possible to associate with each U-structure $\Phi$ on $M^n$ a "negative" complex structure $-\Phi$. Thus given a U-manifold $(M^n, \Phi)$ there is the U-manifold $-(M^n, \Phi) = (M^n, -\Phi)$. It is also possible to associate with each U-structure $\Phi$ on $M^n$ a U-structure $\partial\Phi$ on $\partial M^n$. Define $\partial(M^n, \Phi) = (\partial M^n, \partial\Phi)$. So given a U-manifold $M^n$ it is possible to define $-M^n$ and $\partial M^n$ and these are also U-manifolds. One can then define a bordism relation on closed U-manifolds by $M_1^n \sim M_2^n$ if there exists a compact U-manifold $W^{n+1}$ with $\partial W^{n+1}$ the disjoint union $M_1^n \cup (-M_2^n)$ as U-manifolds. This is an equivalence relation; denote the bordism class containing $M^n$ by $[M^n]$. The set $\Omega_n^U$ of bordism classes $[M^n]$ is an abelian group with addition being disjoint union. Moreover the cartesian product of two U-manifolds is also a U-manifold and $\Omega_*^U = \sum_n \Omega_n^U$ is a graded ring under cartesian product.

We assume the results of Milnor [19] on the structure of $\Omega_*^U$. Namely $\Omega_*^U$ is a polynomial algebra with a generator in each dimension 2k, k > 0. Moreover the Chern numbers $c_{i_1} \cdots c_{i_p}[M^{2n}]$ of any closed U-manifold $M^{2n}$ are functions only of the bordism class $[M^{2n}]$ and also determine the bordism class uniquely as $i_1, \cdots, i_p$ range over all partitions of n.

It follows readily from the results of Milnor that every homomorphism $\varphi: \Omega_{2n}^U \longrightarrow Z$ is a linear combination

$$\varphi[M^{2n}] = \sum a_{i_1 \cdots i_p} c_{i_1} c_{i_2} \cdots c_{i_p} [M^{2n}]$$

with each $a_{i_1 \cdots i_p}$ rational. The problem of Milnor and Hirzebruch can be restated as follows:

PROBLEM: Determine all homomorphisms $\Omega_{2n}^U \longrightarrow Z$.

We shall discuss a solution of Stong [23] and Hattori [15] to this problem in later sections, and go on to further applications.

We shall need a slightly more careful statement of a Milnor result. Given positive integers $i_1 \geq i_2 \geq \cdots \geq i_p$, denote by

$$\sum t_1^{i_1} t_2^{i_2} \cdots t_p^{i_p}$$

the least symmetric polynomial in variables $t_1, \cdots, t_n$ (n large) which contains the term $t_1^{i_1} t_2^{i_2} \cdots t_p^{i_p}$. The symmetric polynomial can be written as a polynomial in the elementary symmetric functions

$$\sum t_1, \sum t_1 t_2, \cdots, t_1 t_2 \cdots t_n.$$

Replace these by $c_1, c_2, \cdots, c_n$ and thus obtain a polynomial

$s_\omega(c_1,\cdots,c_n)$ where $\omega = \{i_1,\cdots,i_p\}$. Given a closed U-manifold $M^{2n}$ and $\omega$ with $i_1 + \cdots + i_p = n$, there is the integer

$$s_\omega[M^{2n}] = \langle s_\omega(c_1(M^{2n}),\cdots,c_n(M^{2n})), \sigma(M^{2n}) \rangle,$$

an integral linear combination of Chern numbers. This number is also denoted by $s_{i_1,\cdots,i_p}[M^{2n}]$. According to Milnor, if 2n is not of the form $2p^k - 2$ for p a prime then there exists a closed U-manifold $M^{2n}$ with $s_n[M^{2n}] = 1$. If $2n = 2p^k - 2$ for p prime, there exists $M^{2n}$ with $s_n[M^{2n}] = p$. Moreover, $\Omega^U_*$ is the polynomial algebra

$$Z[[M^2],[M^4],\cdots,[M^{2n}],\cdots].$$

We put this aside now for later use.

We now take from Chapter I and Chapter II some necessary K-theory for the later sections. First we need the Atiyah classes $\gamma_k$ (see [5]); their existence follows easily from (7.6).

(13.1) <u>There exists a unique function associating with every complex vector space bundle</u> $\xi : E(\xi) \longrightarrow X$, <u>over a finite CW complex, elements</u> $\gamma_k(\xi) \in K(X)$ <u>for</u> $k = 0,1,\cdots$ <u>with</u> $\gamma_0(\xi) = 1$ <u>and such that:</u>

(a) <u>if</u> $\xi, \eta$ <u>are bundles over</u> X,Y <u>respectively and if</u> $f : E(\xi) \longrightarrow E(\eta)$ <u>is a bundle map covering</u> $\overline{f} : X \longrightarrow Y$, <u>then</u> $f^!(\gamma_k(\eta)) = \gamma_k(\xi)$;

(b) <u>if</u> $\xi, \eta$ <u>are bundles over the same space</u> X, <u>then</u> $\gamma_k(\xi + \eta) = \sum_{p+q=k} \gamma_p(\xi)\cdot\gamma_q(\eta)$;

(c) <u>if</u> $\xi$ <u>is a line bundle over</u> X <u>then</u> $\gamma_1(\xi) = \xi - 1$ <u>and</u> $\gamma_k(\xi) = 0$ <u>for</u> $k > 1$.

Of course the $\gamma_k$ are up to sign just the K-theory Chern classes of Chapter II. Just as for cohomology Chern classes, we can form elements $\underline{s}_\omega(\xi) \in K(X)$ for every partition $\omega = \{i_1 \geq i_2 \geq \cdots \geq i_p\}$. Namely instead of the $s_\omega(c_1,\cdots,c_n)$ of a few paragraphs ago, take $\underline{s}_\omega(\xi) = s(\gamma_1, \gamma_2,\cdots, \gamma_n)$. As a special definition denote by $\omega = 0$ the empty partition and let $\underline{s}_\omega(\xi) = 1$. As with ordinary Chern classes, there is the formula

$$\underline{s}_\omega(\xi + \eta) = \sum_{\omega' + \omega''} \underline{s}_{\omega'}(\xi)\cdot \underline{s}_{\omega''}(\eta).$$

If $M^n$ is a U-manifold we may consider the stable tangent bundle $\tau + (2k - n)$ as a complex vector space bundle. The Atiyah classes of this complex vector space bundle are denoted simply by $\gamma_k(M^n) \in K(M^n)$; similarly there are the classes $\underline{s}_\omega(M^n) \in K(M^n)$. The total class $\gamma(M^n)$ is defined to be the formal polynomial $\gamma(M^n) = \sum_0^\infty \gamma_k(M^n)t^k$; more generally there is $\gamma(\xi)$ for any $\xi$.

As an example consider the U-manifold CP(n). Denote by $\xi$ the conjugate $\bar{\rho}$ of the Hopf bundle over CP(n). Then

$$\begin{aligned} \tau + 1 &= (n + 1)\xi \\ \gamma(CP(n)) &= (1 + (\xi - 1)t)^{n+1} \\ &= 1 + \binom{n+1}{1}(\xi - 1)t + \cdots + \binom{n+1}{k}(\xi - 1)^k t^k + \cdots. \end{aligned}$$

Given a U(n)-bundle $\xi : E(\xi) \longrightarrow X$ recall that in Chapter I we have denoted by $\mathcal{J}(\xi)$ a particular Thom class in $K(M(\xi)) = K(D(\xi),S(\xi))$. To make a better fit with standard usage, we shift to a slightly different Thom class

$$T(\xi) = \bar{\mathcal{J}}(\xi) \in \widetilde{K}(M(\xi))$$

where $\bar{\mathcal{J}}$ denotes the complex conjugate of $\mathcal{J}$. We will then use the Thom isomorphism

$$\theta : K(X) \xrightarrow{\simeq} K(D(\xi),S(\xi))$$

sending $x \in K(X)$ into $p^!(x)\cdot T(\xi)$ where $p : D(\xi) \longrightarrow X$ is a bundle projection. More generally, given a finite CW pair $(X,A)$ there is the Thom isomorphism

$$\theta : K(X,A) \xrightarrow{\approx} K(D(\xi),S(\xi) \cup D(\xi|A))$$

sending $x$ into $p^!(x)\cdot T(\xi)$. This is meaningful since

$$p^!(x) \in K(D(\xi),D(\xi)A)),T(\xi) \in K(D(\xi),S(\xi)),$$

and there is the cup product

$$K(Y,B) \otimes K(Y,C) \longrightarrow K(Y,B \cup C).$$

Recall also that there is the ordinary Thom isomorphism

$$\varphi: H^k(X,A) \simeq H^{k+2n}(D(\xi),D(\xi|A) \cup S(\xi))$$

on cohomology; similarly on homology there is

$$\varphi: H_k(X,A) \simeq H_{k+2n}(D(\xi),D(\xi|A) \cup S(\xi)).$$

Denote by $\tilde{H}^{ev}(X,A;Q)$ the commutative ring

$$\tilde{H}^{ev}(X,A;Q) = \Sigma_{k>0} H^{2k}(X,A;Q).$$

Denote by $\tilde{H}^{ev}(X,A;Q)[[t]]$ all formal power series

$$1 + a_1 t + \cdots + a_k t^k + \cdots$$

where $a_k \in H^{ev}(X,A;Q)$. Then $H^{ev}(X,A;Q)[[t]]$ is an abelian group under formal multiplication.

Given a $U(n)$-bundle $\xi: E(\xi) \longrightarrow X$ there is $T(\xi) \in K(D(\xi),S(\xi))$ and

$$\mathrm{ch}\ T(\xi) \in H^{ev}(D(\eta),S(\eta);)Q)[[t]]$$

$$\varphi^{-1}\ \mathrm{ch}\ T(\xi) \in \tilde{H}^{ev}(X;Q)[[t]]$$

Define $T_\xi \in \tilde{H}^{ev}(X;Q)[[t]]$ by

$$T_\xi = \varphi^{-1}\ \mathrm{ch}\ T(\xi).$$

It has been pointed out in Chapter I that

$$T_{\xi+\eta} = T_\xi T_\eta, T_1 = 1.$$

Hence there exists a unique homomorphism $K(X) \longrightarrow H^{ev}(X;Q)[[t]]$ assigning to each $x \in K(X)$ an element $T_x$ and extending the function $\xi \longrightarrow T_\xi$ on bundles.

In particular there is such a homomorphism

$$\tilde{K}(X/A) \rightarrow \tilde{H}^{ev}(X/A;Q)[[t]].$$

Identifying $\tilde{K}(X/A)$ with $K(X,A)$ and $\tilde{H}^{ev}(X/A;Q)$ with $\tilde{H}^{ev}(X,A;Q)$ we thus get a homomorphism

$$T : K(X,A) \longrightarrow \tilde{H}^{ev}(X,A;Q)[[t]]$$

assigning to each $x \in K(X,A)$ an element $T_x$.

In an entirely similar way given $\xi$, there are the elements

$$c(\xi) = \Sigma c_k(\xi)t^k,\quad \gamma(\xi) = \Sigma \gamma_k(\xi)\, t^k$$

in $\tilde{H}^{ev}(X;Q)[[t]]$. Just as above these lead to homomorphisms

$$c : K(X,A) \longrightarrow \tilde{H}^{ev}(X,A;Q)[[t]]$$

$$\gamma : K(X,A) \longrightarrow K(X,A)[[t]]$$

mapping x into

$$c(x) = 1 + c_1(x)t + \cdots + c_k(x)t^k + \cdots$$

$$\gamma(x) = 1 + \gamma_1(x)t + \cdots + \gamma_k(x)t^k + \cdots.$$

The element $(T_x)^{-1} \in \tilde{H}^{ev}(X,A;Q)[[t]]$ will be called the Todd polynomial of x. Letting $c_k = c_k(x)$, it is the famous polynomial

$$(T_x)^{-1} = 1 + \frac{c_1}{2}t + \frac{c_1^2 + c_2}{12}t^2 + \frac{c_1c_2}{24}t^3 + \cdots$$

of Hirzebruch [16].

Suppose now that $M^n$ is a U-manifold. The stable tangent bundle $\tau + (2k - n)$ is then a complex vector space bundle, unique up to equivalence. Denote by $\tau' = \tau'(M) \in K(M^n)$ the element represented by this complex bundle. Then let

$$T(M^n) = T(\tau'), T^{-1}(M^n) = [T(\tau')]^{-1}.$$

Then $T^{-1}(M^n)$ is the Todd polynomial of the U-manifold $M^n$.

It should be noted from (6.4) that the precise value of $T_\xi$ is given as follows: namely in

$$\frac{(1 - e^{-t_1})\cdots(1 - e^{-t_n})}{t_1\cdots t_n}$$

replace $\Sigma t_1$ by $c_1(\xi), \cdots, \Sigma t_1 \cdots t_k$ by $c_k(\xi), \cdots$.

13. Characteristic numbers from K-theory.

Assume that we are given a compact U-manifold $M^{2n}$ and an element $x \in K(M, \partial M)$. Embed M as a smooth submanifold of the cube $I^{2n+2k}$, $2k \geq 2n + 2$, so that $M \cap \partial I^{2n+2k} = \partial M$, so that this intersection is in the interior of one face of $I^{2n+2k}$ and so that M is orthogonal to $\partial I^{2n+2k}$ at this intersection. Denote by $\eta$ the normal bundle to M in $I^{2n+2k}$. Since $M^{2n}$ is a U-manifold, its stable tangent bundle $\tau'$ is a complex vector space bundle. We can then suppose that the normal bundle $\eta$ to M in $I^{2n+2k}$ is a complex vector space bundle with $\tau' + \eta$ trivial (see [12, p.16]). Then $T_\eta = T^{-1}_{\tau'} = T^{-1}(M)$. The disk bundle $D(\eta)$ may be identified with the tubular neighborhood of $\partial M$ in $\partial I^{2n+2k}$. We then have the Thom class

$$T(\eta) \in K(D(\eta), S(\eta)), \text{ and}$$
$$p^!(\eta) \in K(D(\eta), D(\eta|\partial M))$$

where $p : (D(\eta), D(\eta|\partial M)) \longrightarrow (M, \partial M)$ is bundle projection. Using the cup product

$$K(X,A) \otimes K(X,B) \longrightarrow K(X, A \cup B)$$

we then get

$$\theta : K(M, \partial M) \approx K(D(\eta), S(\eta) \cup D(\eta|\partial M))$$

sending x into $(p^! x)\cdot T(\eta)$. There is the diagram

$$\begin{array}{ccc} K(M, \partial M) & & K(I^{2n+2k}, \partial I^{2n+2k}) \approx Z \\ \theta\downarrow\approx & & \uparrow j^! \\ K(D(\eta), S(\eta) \cup D(\eta|\partial M)) & \underset{\approx}{\overset{i^!}{\longleftarrow}} & K(I^{2n+2k}, \partial I^{2n+2k} \cup (I^{2n+2k} - \operatorname{Int} D(\eta))). \end{array}$$

By definition, given the compact U-manifold $M^{2n}$ and the element $x \in K(M, \partial M)$, denote by $x[M]$ the integer which is the image of x under the composite homomorphism $K(M, \partial M) \longrightarrow Z$.

(13.1) <u>Given a compact U-manifold M and $x \in K(M, \partial M)$, the integer $x[M]$ is given by</u>

$$x[M] = \langle ch\, x\cdot T^{-1}(M), \sigma(M) \rangle .$$

<u>In particular it is independent of the choices made.</u>

Proof. We have

$$\begin{aligned} x[M] &= \langle \mathrm{ch}\, j^{!}(i^{!})^{-1}\theta(x),\ \sigma(I^{2n+2k})\rangle \\ &= \langle \mathrm{ch}\,\theta(x),\ \sigma(D(\eta))\rangle \\ &= \langle \varphi^{-1}\,\mathrm{ch}\,\theta(x),\ \sigma(M)\rangle \\ &= \langle \varphi^{-1}(\mathrm{ch}\, p^{!}x\cdot \mathrm{ch}\, T(\eta)),\ \sigma(M)\rangle \\ &= \langle \mathrm{ch}\, x\cdot \varphi^{-1}\mathrm{ch}\, T(\eta),\ \sigma(M)\rangle \\ &= \langle \mathrm{ch}\, x\cdot T^{-1}(M),\ \sigma(M)\rangle . \end{aligned}$$

The remark follows.

In particular if $M^{2n}$ is a closed U-manifold and $x = 1$, then

$$x[M^{2n}] = \langle T^{-1}(M^{2n}),\ \sigma(M^{2n})\rangle$$

is the Todd index of $M^{2n}$. <u>We denote it by</u> Td $[M^{2n}]$.

As an example, consider the integers $x[CP(n)]$ where $x$ ranges over the elements of $K(CP(n))$. Note that

$$T^{-1}(CP(n)) = (t/(1 - e^{-t}))^{n+1}$$

where $t$ is the appropriate generator of $H^2(CP(n))$. We assume the fact (see Hirzebruch [16]) that the coefficient of $t^n$ in the above is one, and thus that Td $[CP(n)] = 1$.

Let $x = (1 - \rho)^k$ where $\rho$ is the Hopf bundle and $0 \leq k \leq n$. Then

$$\begin{aligned} \mathrm{ch}\, x\cdot T^{-1}(CP(n)) &= (1 - e^{-t})^k (t/(1 - e^{-t}))^{n+1} \\ &= t^k (t/(1 - e^{-t}))^{n-k+1}. \end{aligned}$$

It follows that the coefficient of $t^n$ in $\mathrm{ch}\, x\cdot T^{-1}(CP(n))$ is 1, thus

$$(1 - \rho)^k[CP(n)] = 1,\ 0 \leq k \leq n.$$

(13.2) Let $x \in K(CP(n))$. Then

$$x = a_o + a_1(1 - \rho) + \cdots + a_n(1 - \rho)^n$$

for integers $a_o, a_1, \cdots, a_n$ and

$$x[CP(n)] = a_o + a_1 + \cdots + a_n.$$

Consider the conjugate $\bar{\rho}$ of $\rho$. Then $\rho\bar{\rho} = 1$ and

$$\begin{aligned} \bar{\rho} &= 1/1 - (1 - \rho) \\ &= 1 + (1 - \rho) + (1 - \rho)^2 + \cdots \\ \bar{\rho} - 1 &= (1 - \rho) + (1 - \rho)^2 + \cdots \end{aligned}$$

(13.3) The number $(\bar{\rho} - 1)^k[CP(n)]$ is equal to $\binom{n}{k}$.

Proof. We have only to expand $(\bar{\rho} - 1)^k$ by the above formula and use (13.2). We see that the result is the number of sequences $i_1, \cdots, i_k$ with $1 \leq i_j \leq n$ and with $i_1 + \cdots + i_k \leq n$. Using induction, we see that the remark is implied by the identity

$$\binom{n}{k} = \binom{n-1}{k-1} + \binom{n-2}{k-1} + \cdots + \binom{k-1}{k-1}.$$

Suppose now that $M^n$ is a closed U-manifold and that $c \in H^2(M^n)$. Since $K(Z,2) = CP(\infty)$, there is a map $f : M \longrightarrow CP(N)$, $N$ large, with $f^*(t) = c$. Since $CP(\infty) = BU(1)$, we see that there exists a complex line bundle $\eta$ on $M$ with $c_1(\eta) = c$. Making $f$ transverse regular on $CP(N - 1)$, denote by $N^{n-2}$ the inverse image $f^{-1}(CP(N - 1))$. By [12, p. 16], we can make $N^{n-2}$ into a U-manifold whose stable tangent bundle $\tau'$ has $\tau' + i^*\eta = i^*\tau'(M^n)$, where $i : N^{n-2} \subset M^n$. We call $N^{n-2}$ a U-submanifold dual to $c \in H^2(M^n)$.

(13.4) Suppose that $M^n$ is a closed U-manifold and that $c \in H^2(M^n)$. Denote by $\eta$ a complex line bundle on $X$ with $c_1(\eta) = c$

*and by* $N^{n-2} \subset M^n$ *a* U-*submanifold dual to* c. *If* $x \in K(M^n)$ *and* $i : N^{n-2} \subset M^n$ *then*

$$(i^! x)[N^{n-2}] = x(1 - \bar{\eta})[M^n].$$

Proof. Since $\tau'(M) + i^*\eta = i^*\tau'(M)$, we have

$$T^{-1}(N)/T_{i^*\eta} = i^*T^{-1}(M)$$

$$T^{-1}(N) = i^*(T_\eta \cdot T^{-1}(M)).$$

Hence

$$\begin{aligned}(i^! x)[N^{n-2}] &= \langle ch\ i^! x \cdot T^{-1}(N), \sigma(N)\rangle \\ &= \langle i^*(ch\ x \cdot T_\eta \cdot T^{-1}(M), \sigma(N)\rangle \\ &= \langle ch\ x \cdot T_\eta \cdot T^{-1}(M), i_* \sigma(N)\rangle \\ &= \langle ch\ x \cdot T_\eta \cdot T^{-1}(M), c \cap \sigma(M)\rangle \\ &= \langle ch\ x \cdot cT_\eta \cdot T^{-1}(M), \sigma(M)\rangle \\ &= \langle ch\ x(1 - e^{-c})T^{-1}(M), \sigma(M)\rangle \\ &= x(1 - \bar{\eta})[M^n].\end{aligned}$$

14. *The theorem of Stong and Hattori.*

We have mentioned in section 12 the standard invariants $s_{i_1,\ldots,i_k}[M^{2n}]$ of a closed U-manifold $M^{2n}$, defined whenever $i_1 + \cdots + i_k = n$. Using section 13 we now have the integers

$$\underline{s}_{i_1,\ldots,i_k}[M^{2n}] = \underline{s}_{i_1,\ldots,i_k}(M^{2n})[M^{2n}]$$

defined whenever $i_1 + \cdots + i_k \leq n$. If $i_1 + \cdots + i_k = n$ it is readily checked that

$$\underline{s}_{i_1,\ldots,i_k}[M^{2n}] = s_{i_1,\ldots,i_k}[M^{2n}].$$

It follows readily from (13.1) that the $\underline{s}_{i_1,\cdots,i_k}[M^{2n}]$ are rational combinations of the Chern numbers, thus they are bordism invariants of closed U-manifolds. Hence we receive homomorphisms

$$\underline{s}_{i_1,\cdots,i_k} : \Omega^U_{2n} \longrightarrow Z,$$

defined whenever $i_1 + \cdots + i_k \leq n$. For the empty partition $\omega = 0$, we have

$$\underline{s}_0[M^{2n}] = \langle T^{-1}(M^{2n}), \sigma(M^{2n})\rangle = \mathrm{Td}\,[M^{2n}].$$

(14.1) <u>If</u> $\omega = \{i_1, i_2, \cdots, i_k\}$ <u>is a partition consisting of</u> 1 <u>repeated</u> $r_1$ <u>times</u>, 2 <u>repeated</u> $r_2$ <u>times</u>, $\cdots$, s <u>repeated</u> $r_s$ <u>times, then</u>

$$\underline{s}_\omega[CP(n)] = \frac{(n+1)!}{r_1!\cdots r_s!(n+1-r_1-\cdots-r_s)} \cdot \binom{n}{i_1 + \cdots + i_k}.$$

Proof. The tangent bundle $\tau$ of CP(n) is given by

$$\tau + 1 = (n+1)\,\xi, \quad \xi = \bar{\gamma}$$

hence

$$\underline{s}_\omega(\tau) = \sum \underline{s}_{i_1}(\xi)\cdots\underline{s}_{i_k}(\xi),$$

there being $(n+1)!/r_1!\cdots r_s!(n+1-r_1-\cdots-r_s)!$ such terms. Now

$$\underline{s}_{i_1}(\xi)\cdots s_{i_k}(\xi) = (\xi - 1)^{i_1 + \cdots + i_k},$$

and

$$(\xi - 1)^{i_1 + \cdots + i_k}[CP(n)] = \binom{n}{i_1 + \cdots + i_k}.$$

The remark follows.

We have next a fundamental computation of Stong [23].

(14.2) <u>Consider</u> $(CP(p^k))^p = CP(p^k) \times \cdots \times CP(p^k)$, <u>for</u> p <u>a given</u>

*prime, and let* $c \in H^2((CP(p^k))^p)$ *be given by*

$$c = t\otimes 1\otimes \cdots \otimes 1 + 1\otimes t \otimes 1\otimes \cdots \otimes 1 + \cdots + 1\otimes 1\otimes \cdots \otimes 1\otimes t,$$

*where* $t$ *is the preferred generator of* $H^2(CP(p^k))$. *Let* $N = N^{2p^{k+1}-2}$ *be be the U-submanifold of* $(CP(p^k))^p$ *dual to* $c$. *Then*

i) $\underline{s}_{i_1,\cdots,i_\chi}[N] = 0$ *mod* $p$, $i_1 + \cdots + i_\chi > p^{k+1} - p$

*If* $i_1 + \cdots + i_\chi = p^{k+1} - p$, *then*

ii) $\underline{s}_{i_1,\cdots,i_\chi} = 0$ *mod* $p$, $\chi < p$

iii) $\underline{s}_{i_1,\cdots,i_p} = 0$ *mod* $p$ unless $i_j = p^k - 1$ *for each* $j$

iv) $\underline{s}_{\underbrace{p^k-1,\cdots,p^k-1}_{p \text{ terms}}} = 1$ *mod* $p$.

Proof. Let $M = (CP(p^k))^p$. Let $x_1,\cdots,x_p \in K(M)$ be given by

$$x_i = 1\otimes \cdots \otimes 1\otimes (\xi - 1) \otimes 1\otimes \cdots \otimes 1$$

where $\xi - 1$ is in $i$-th position and where $\xi = \bar{\rho}$. Then $K(M)$ consists of all polynomials

$$\varphi(x_1,\cdots,x_p) = \sum a_{r_1,\cdots,r_p} x_1^{r_1}\cdots x_p^{r_p}$$

for which each $r_i \leq p^k$. Moreover

$$x_1^{r_1}\cdots x_p^{r_p}[M] = (\xi - 1)^{r_1}[CP(p^k)]\cdots(\xi - 1)^{r_p}[CP(p^k)]$$

which is, by (13.3), 0 mod p unless each $r_i$ is either 0 or $p^k$.

We will be especially interested in the symmetric polynomials, therefore for $\omega = \{r_1 \geq r_2 \geq \cdots \geq r_p\}$, $r_1 \leq p^k$, in the symmetric polynomials

$$x_\omega = \sum x_1^{r_1} x_2^{r_2}\cdots x_p^{r_p}.$$

Note that $x_\omega[M] = 0$ mod p unless $\omega = 0$ or $\omega = (p^k,\cdots,p^k)$.

For each $\omega$, let

$$x^*_\omega = \begin{cases} 0 \text{ unless } r_1 = \cdots = r_p \\ x_\omega \text{ if } r_1 = \cdots = r_p. \end{cases}$$

Extend $x_\omega \longrightarrow x^*_\omega$ to a homomorphism assigning to each symmetric polynomial $\varphi$ a symmetric polynomial $\varphi^*$. Note that $(xy)^* = x^*y^*$ mod p for x,y symmetric. Adding this to the previous fact that $x_\omega[M] = 0$ unless $\omega = 0$ or $\omega = (p^k,\cdots,p^k)$ we get

$$\begin{aligned}(xy\cdots w)[M] &= (xy\cdots w)^*[M] \bmod p \\ &= x^*y^*\cdots w^*[M] \bmod p\end{aligned}$$

for symmetric $x,y,\cdots,w \in K(M)$.

Consider in particular the line bundle $\eta$ over M with $c_1(\eta) = c$. Then

$$\begin{aligned}\eta &= \xi \otimes \xi \otimes \cdots \otimes \xi \\ &= (1 + (\xi - 1)) \otimes \cdots \otimes (1 + (\xi - 1)) \\ &= 1 + \sum x_1 + \sum x_1x_2 + \cdots + x_1x_2\cdots x_p \\ (\eta - 1)^* &= x_1x_2\cdots x_p.\end{aligned}$$

Also $(1 - \bar{\eta}) = (\eta - 1) + (\eta - 1)^2 + \cdots$

$$(1 - \bar{\eta})^* = x_1\cdots x_p + x_1^2\cdots x_p^2 + \cdots \bmod p$$

Note for later reference that $[(\eta - 1)^*(1 - \bar{\eta})]^*$ mod p has term of lowest degree $x_1^{j+1}\cdots x_p^{j+1}$. Hence if x is a homogeneous polynomial of degree q and $q + j \geq p^{k+1} - p$, where $j > 0$, then

$$\begin{aligned}x(\eta - 1)^j(1 - \bar{\eta})[M] &= x^*(\eta - 1)^{*j}(1 - \bar{\eta})^*[M] \bmod p \\ &= 0 \bmod p\end{aligned}$$

if either $j > 0$ or if $q > p^{k+1} - p$ and $j = 0$. This is true because $x^*(\eta - 1)^{*j}(1 - \bar{\eta})^*$ is of degree $> p^{k+1}$.

Let $i : N \subset M$ and let $\eta' = i^*\eta$. Then $i^*\tau'(M) = \tau(N) + \eta'$ and

$$i^*\underline{s}_\omega(M) = \underline{s}_\omega(N) + \sum \underline{s}_{i_1,\ldots,i_{j-1},i_{j+1},\ldots,i_\chi}(N)\cdot(\eta' - 1)^{i_j}$$

$$\underline{s}_\omega(N) = i^*\underline{s}_\omega(M) - \sum \underline{s}_{i_1,\ldots,i_{j-1},i_{j+1},\ldots,i_\chi}(N)(\eta' - 1)^{i_j}.$$

Proceeding inductively, we see that

$$\underline{s}_\omega(N) = i^*(\underline{s}_\omega(M) + \sum a_{\omega'} x_{\omega'}(\eta' - 1)^{j_{\omega'}})$$

where the $a_{\omega'}$ are integers, the $x_{\omega'}$ are homogeneous polynomials and $\deg x_{\omega'} + j_{\omega'} = \deg \omega$. Also $j_{\omega'} > 0$. Here we use the fact that $\underline{s}_\omega(M)$ is homogeneous of degree $\deg \omega$ in $x_1,\cdots,x_p$.

Let $\omega$ be a partition of degree $\geq p^{k+1} - p$. Then

$$\underline{s}_\omega[N] = \underline{s}(M)(1 - \bar{\eta})[M] + \sum a_{\omega'} x_{\omega'}(\eta' - 1)^{j_{\omega'}}(1 - \bar{\eta})[M]$$

By an earlier computation we then get

$$\underline{s}_\omega(N) = \underline{s}(M)(1 - \bar{\eta})[M] \bmod p$$

for $\deg \geq p^{k+1} - p$. Also by our earlier computation we get

$$\underline{s}(M)(1 - \bar{\eta})[M] = 0 \bmod p, \ \deg \ > p^{k+1} - p.$$

Hence i) is established.

Suppose next that $\omega = \{i_1 \geq \ldots \geq i_\chi\}$, with $i_1 + \cdots + i_\chi = p^{k+1} - p$. Then

$$\underline{s}_\omega[N] = (\underline{s}_\omega(M)^*(x_1\cdots x_p)[M] \bmod p.$$

If $\chi < p$ then

$$\underline{s}_\omega[M] = \Sigma \underline{s}_{i_1}[CP(p^k)] \otimes \cdots \otimes s_{i_\chi}[CP(p^k)] \otimes 1 \otimes \cdots \otimes 1$$

clearly has $(\underline{s}_\omega[M])^* = 0$ and hence ii) follows. Similarly if $\chi = p$ then $(\underline{s}_\omega(M))^* = 0$ unless $i_1 = \cdots = i_p$ and hence iii) follows. If $i_1 = \cdots = i_p = p^k - 1$ then

$$\begin{aligned} \underline{s}_\omega[N] &= s_\omega(M)^*(1 - \bar{\eta})^*[M] \bmod p \\ &= x_1^{p^k} \cdots x_p^{p^k}[M] \bmod p \\ &= 1 \bmod p. \end{aligned}$$

Both Stong [23] and Hattori [15] have given proofs of the following theorem. We are using Stong's proof here.

(14.3) STONG. <u>Suppose that</u> $\varphi: \Omega^U_{2n} \longrightarrow Z$ <u>is a homomorphism. Then</u> $\varphi$ <u>can be expressed as an integral linear combination of the homomorphism</u>

$$\underline{s}_{i_1,\ldots,i_k} : \Omega^U_{2n} \longrightarrow Z, \; i_1 + \cdots + i_k \leq n.$$

Proof. Given a partition $\omega = \{i_1 \geq i_2 \geq \cdots \geq i_k\}$, let let $d(\omega) = i_1 + \cdots + i_k$ and $n(\omega) = k$. Suppose that $\omega' = \{j_1 \geq \cdots \geq j_\chi\}$. Define

$\omega' > \omega$ if $d(\omega') > d(\omega)$,

$\omega' > \omega$ if $d(\omega') = d(\omega)$ and $n(\omega') < n(\omega)$,

$\omega' > \omega$ if $d(\omega') = d(\omega), n(\omega') = n(\omega)$

and if $j_1 = i_1, \ldots, j_s = i_s, j_{s+1} > i_{s+1}$. This is a linear ordering of all partitions. We leave it as an exercise to show that if $\omega_1 \leq \omega_1'$ and $\omega_2 \leq \omega_2'$ then $\omega_1 + \omega_2 \leq \omega_1' + \omega_2'$.

Let $[M] \in \Omega^U_*$. Say that $[M]$ is of type $\omega$ if $\underline{s}_\omega[M] \neq 0 \bmod p$ and if $\underline{s}_{\omega'}[M] = 0 \bmod p$ for all $\omega' > \omega$. If $M$ is of type $\omega_1$ and if $N$ if of type $\omega_2$ then $M \times N$ is of type $\omega_1 + \omega_2$. For instance, suppose

$\omega > \omega_1 + \omega_2$. Then

$$\underline{s}_{\omega}[M \times N] = \sum_{\omega' + \omega'' = \omega} \underline{s}_{\omega'}[M] \underline{s}_{\omega''}[N].$$

If $\omega' + \omega'' = \omega > \omega_1 + \omega_2$ then clearly either $\omega' > \omega_1$ or $\omega'' > \omega_2$. Hence

$$s_{\omega}[M \times N] \equiv 0 \bmod p, \ \omega > \omega_1 + \omega_2.$$

Similarly $\underline{s}_{\omega_1 + \omega_2}[M \times N] \not\equiv 0 \bmod p$.

In each positive dimension 2k, we now select a closed U-manifold $X^{2k}$. If $k \neq p^i - 1$, for p a given prime, let $X^{2k}$ be such that

$$s_k[X^{2k}] = \underline{s}_k[X^{2k}] = 1.$$

Then $[X^{2k}]$ is of type $\omega(k) = k$.

If $k = p - 1$, let $X^{2k} = CP(p - 1)$. It follows readily from (15.1) that $[X^{2p-2}]$ is of type $\omega(\gamma - 1) = 0$.

If $k = p^{r+1} - 1$, $(n \geq 1)$ let $X^{2k}$ be the U-submanifold of $(CP(p^r))^p$ dual to c as in (14.2). According to (15.2), $[X^{2p^{r+1}-2}]$ is of type $(p^{r+1} - 1) = (p^r - 1, \ldots, p^r - 1)$.

Fix now the positive integer n, and consider $\Omega^U_{2n} \otimes Z_p = \Omega^U_{2n}/p\,\Omega^U_{2n}$. Consider partitions $\{i_1, \cdots, i_k\}$ of n. For each such we have

$$M^{2i_1} \times \cdots \times M^{2i_k} \in \Omega^U_{2n}/p\,\Omega^U_{2n}$$

and the partition

$$\omega(i_1, \ldots, i_k) = \omega(i_1) + \ldots + \omega(i_k)$$

of degree $\leq n$. Moreover $[M^{2i_1}] \times \cdots \times [M^{2i_k}]$ is of type $\omega(i_1, \cdots, i_k)$.

If $[X] = \sum a_{i_1,\dots,i_k}[M^{2i_1}\times\dots\times M^{2i_k}] \in \Omega^U_{2n}/p\,\Omega^U_{2n}$

with some $a_{i_1,\dots,i_k} \neq 0 \bmod p$, summed over all $\{i_1,\dots,i_k\}$ with $i_1+\dots+i_k = n$, let $a_{j_1,\dots,j_\chi}$ be such that $\omega(j_1,\dots,j_\chi)$ takes its maximum. Then [X] is of type $\omega(j_1,\dots,j_\chi)$.

That is, let $(Z_p)^{\pi(n)}$ be a direct product of $\pi(n)$ copies of $Z_p$ ($\pi(n)$ = number of partitions of n), and let it be indexed by the partitions $\{i_1 \geq \dots \geq i_k\}$ of n. Define

$$\Phi : \Omega^U_{2n}/p\,\Omega^U_{2n} \longrightarrow (Z_p)^{\pi(n)}$$

by

$$\Phi[M] = (\underline{s}_{\omega(i_1,\dots,i_k)}[M]).$$

Then $\Phi$ is an isomorphism, as follows from the above paragraph.

It follows readily that every homomorphism $\theta : \Omega^U_{2n} \longrightarrow Z_p$ is equal, mod p, to an integral linear combination of the $\underline{s}_{\omega(i_1,\dots,i_k)}$. In particular, every $\theta$ is equal mod p to an integral linear combination of the $\underline{s}_\omega$ as $\omega$ varies over all partitions.

Consider the free abelian group Hom $(\Omega^U_{2n}, Z)$ of rank $\pi(n)$, and let

$$K \subset \text{Hom}\,(\Omega^U_{2n}, Z) = G$$

be the subgroup spanned by all the $\underline{s}_\omega$ as $\omega$ varies over all partitions with $d(\omega) \leq n$. Now G/K is clearly a finite abelian group. Hence there exists a basis $\varphi_1,\dots,\varphi_{\pi(n)}$ for G and positive integers $r_1,\dots,r_{\pi(n)}$ so that

$$r_1\varphi_1,\dots,r_{\pi(n)}\varphi_{\pi(n)}$$

is a basis for K. Suppose $r_j > 1$. Some prime p is then a divisor of $r_j$. Then $K \otimes Z_p \neq G \otimes Z_p$. But from the preceding part of the proof we have

$$K \otimes Z_p = G \otimes Z_p = \mathrm{Hom}\,(\Omega^U_{2n}, Z_p).$$

The theorem follows.

(14.4) COROLLARY. Recall that for each partition $\omega = \{i_1, \ldots, i_k\}$ with $d(\omega) \leq n$ the integer $s_\omega[M^{2n}] = \sum r_{j_1 \ldots j_\chi} c_{j_1} c_{j_r} \cdots c_{j_\chi}[M^{2n}]$ where the r's are rational and $j_1 \geq \cdots \geq j_\chi$ varies over all partitions of n. Given integers $a_{j_1, \ldots, j_\chi}$, one for each partition of n, let $a = (a_{j_1, \ldots, j_\chi} : j_1 + \cdots + j_\chi = n)$ and let

$$s_\omega(a) = \sum r_{j_1 \ldots j_\chi}\ a_{j_1 \ldots j_\chi}.$$

A necessary and sufficient condition that there exist a closed U-manifold $N^{2n}$ with $c_{j_1} c_{j_2} \cdots c_{j_\chi}[N^{2n}] = a_{j_1, \ldots, j_\chi}$ for each $\{j_1, \ldots, j_\chi\}$ is that $s_\omega(a)$ be an integer for all partitions $\omega$ with $d(\omega) \leq n$.

Proof. It is clear that this is a necessary condition. We have only to prove it sufficient. Since $\Omega^U_{2n} \approx (Z)^{\pi(n)}$, it follows from (14.3) that there exists integral linear combinations $\varphi_1, \ldots, \varphi_{\pi(n)}$ of the $s_\omega$ such that

$$\rho : [M^{2n}] \longrightarrow (\varphi_1[M^{2n}], \ldots, \varphi_{\pi(n)}[M^{2n}])$$

is an isomorphism $\Omega^U_{2n} \approx (Z)^{\pi(n)}$. There is also the embedding $\Omega^U_{2n} \subset (Z)^{\pi(n)}$ using Chern numbers, namely $[M^{2n}]$ is identified with the set of $\pi(n)$ Chern numbers $c_{j_1} \ldots c_{j_\chi}[M^{2n}]$, ordered in some way. Finally there is

$$\rho' : (Z)^{\pi(n)} \longrightarrow (Q)^{\pi(n)}$$

sending $a = (a_{j_1},\dots,a_{j_\chi})$ into $(\varphi_1(a),\dots,\varphi_{\pi(n)}(a))$. Commutativity holds in

$$\begin{array}{ccc} \Omega^U_{2n} & \xrightarrow[\approx]{\rho} & (Z)^{\pi(n)} \\ \cap & & \cap \\ (Z)^{\pi(n)} & \xrightarrow{\rho'} & (Q)^{\pi(n)} . \end{array}$$

Moreover $(Z)^{\pi(n)}/\Omega^U_{2n}$ is of finite order. Hence $\rho'$ is a monomorphism and the corollary follows readily.

15. U-manifolds with stably framed boundaries.

Let $M^n$ denote a differentiable manifold and let $\tau$ denote its tangent bundle. Denote as usual the stable tangent bundle of $M^n$ to be $\tau + (2k - n$ where $2k - n \geq 2$. A stable framing $\theta$ of $M^n$ is a homotopy class of maps

$$\varphi: E(\tau + (2k - n)) \longrightarrow R^{2k}$$

each of which maps every fiber of $\tau + (2k - n)$ linearly onto $R^{2k}$. As with U-structures, this is independent of the value of k as long as $2k - n \geq 2$ (see [12, p. 16]).

A stably framed manifold is a pair $(M^n,\theta)$ consisting of a differentiable manifold $M^n$ and a stable framing $\theta$ of $M^n$.

There is a bordism group $\Omega^{fr}_n$ of bordism classes of stably framed closed manifolds. As with U-structures, given a stable framing $\theta$ of $M^n$ one can define a stable framing $-\theta$; one can also define a stable framing $\partial\theta$ on $\partial M^n$ and thus define

$$-(M^n,\theta) = (M^n,-\theta), \quad \partial(M^n,\theta) = (\partial M^n, \partial\theta).$$

For more details see [12]. One can then define a bordism relation on closed stably framed manifolds by $M_1^n \sim M_2^n$ if there exists a compact stably framed manifold $W^{n+1}$ with $\partial W^{n+1}$ the disjoint union $M_1^n \cup (-M_2^n)$ as stably framed manifolds. Denote the bordism class containing $M^n$ by $[M^n]_{fr}$ and denote the abelian group of bordism classes by $\Omega_n^{fr}$. The cartesian product of two stably framed manifolds is stably framed and $\Omega_*^{fr} = \Sigma_n \Omega_n^{fr}$ is a graded ring under cartesian product [12].

The abelian group $\Omega_n^{fr}$ is known to be isomorphic to the stable stem $\pi_{n+2k}(S^{2k})$, $2k \geq n + 2$, by the method of Thom. In particular $\Omega_0^{fr} \approx Z$ and $\Omega_n^{fr}$ is finite for $n > 0$.

Every stable framing $\theta$ on $M^n$ gives rise to a U-structure on $M^n$. For given

$$\varphi : E(\tau + (2k - n)) \longrightarrow R^{2k}$$

the natural operator $J : R^{2k} \longrightarrow R^{2k}$ given by

$$J(x_1, x_2, \cdots, x_{2n-1}, x_{2k}) = (-x_2, x_1, \cdots, -x_{2k}, x_{2k-1})$$

pulls back to an operator

$$J : E(\tau + (2k - n)) \longrightarrow E(\tau + (2k - n))$$

representing a U-structure $\theta'$ on $M^n$. This leads to a homomorphism

$$r : \Omega_n^{fr} \longrightarrow \Omega_n^U$$

mapping $[M^n, \theta]_{fr}$ into $[M^n, \theta']_U$. For $n > 0$, $\Omega_n^{fr}$ is finite and $\Omega_n^U$ is free abelian. Hence $r = 0$ for $n > 0$.

That is, given a closed stably framed manifold $M^n$, $n > 0$, then

$M^n$ is a U-manifold and $[M^n] = 0$ in $\Omega^U_n$. Hence there exists a compact U-manifold $W^{n+1}$ with $\partial W^{n+1} = M^n$. The point of the remainder of this chapter is to consider such pairs $(W^{n+1}, M^n)$.

A (U,fr)-<u>manifold</u> is a triple $(M^n, \Phi, \theta)$ consisting of a differentiable manifold $M^n$, a U-structure $\Phi$ on $M^n$ and a stable framing $\theta$ of $\partial M^n$ such that $\theta' = \partial \Phi$. Many non-trivial examples exist by virtue of the above construction. Picking representatives of $\Phi$ and $\theta$, we may regard the stable tangent bundle $\tau + (2k - n)$ as a complex vector space bundle on $M^n$ with a given trivialization, as a complex vector space bundle, when restricted to $\partial M^n$.

Denote by $\tau'$ the stable tangent bundle of $M^n$, a bundle of k-dimensional complex vector spaces. Moreover we are given an isomorphism $\varphi$ of $\tau' | \partial M^n$ with the trivial bundle k on $\partial M^n$. The difference class

$$d(\tau', k, \varphi) \in K(M^n, \partial M^n)$$

will be called the stable tangent bundle of the (U,fr)-manifold $M^n$. It is independent of the various choices made. Define

$$\tau = \tau(M, \partial M) \in k(M, \partial M)$$

to be this element. The (U,fr)-manifold M then has Chern classes $c_k(M) = c_k(\tau)$ in $H^{2k}(M, \partial M)$. Therefore we can define Chern numbers of a compact (U,fr)-manifold by

$$c_{i_1} \cdots c_{i_\chi}[M^{2n}] = \langle c_{i_1}(M) \cdots c_{i_\chi}(M), \sigma(M) \rangle .$$

The main purpose of this chapter is to solve the following.

PROBLEM. Given a compact (U,fr)-manifold $M^{2n}$, when is there a closed U-manifold $N^{2n}$ having the same Chern numbers? That is, to what extent can $[\partial M^{2n}]_{fr} \in \Omega^{fr}_{2n-1}$ be detected by the Chern numbers of $M^{2n}$?

Note that the Todd genus of a compact (U,fr)-manifold is defined as a rational number. For closed U-manifolds, there is a rational linear combination of Chern numbers giving $Td[M^{2n}]$. Simply use this to define $Td[M^{2n}]$ for $M^{2n}$ a compact (U,fr)-manifold. More precisely, since $\tau \in K(M, \partial M)$ there is $T(\tau) \in \tilde{H}^{ev}(M, \partial M; Q)[[t]]$, also $T^{-1}(M) = (T(\tau))^{-1}$ in $\tilde{H}^{ev}(M, \partial M; Q)[[t]]$, and

$$Td[M^{2n}] = \langle T^{-1}(M), \sigma(M) \rangle .$$

We give now a few examples. Let $M^n$ be a compact differentiable oriented manifold and let $x \in K(M^n, \partial M^n)$ be such that the composition

$$K(M^n, \partial M^n) \longrightarrow \tilde{K}(M^n) \rightarrow KO(M^n)$$

maps x into the class of the stable tangent bundle in $KO(M^n)$. We may then use x to put a complex vector space structure on the stable tangent bundle and one with a trivialization on its restriction to the boundary. That is, we can make $M^n$ into a (U,fr)-manifold (not uniquely) with $\tau(M, \partial M) = x \in K(M, \partial M)$. Moreover $c_k(M) = c_k(x)$.

Thus on the 2n-disk $D^{2n}$, let $x \in K(D^{2n}, S^{2n-1})$ have $\langle c_n(x), \sigma(D^{2n}) \rangle = (n-1)!$ Since $\widetilde{KO}(D^{2n}) = 0$, we may consider $D^{2n}$ a (U,fr)-manifold with $\tau = x$. Thus there exists a compact (U,fr)-manifold $D^{2n}$ with

$$c_n[D^{2n}] = (n-1)!$$

and all other Chern numbers zero. Then

$$\mathrm{Td}[D^2] = c_1[D^2]/2 = 1/2$$
$$\mathrm{Td}[D^4] = c_2[D^4]/12 = 1/12$$
$$\mathrm{Td}[D^8] = c_4[D^4]/720 = 1/120$$

etc.

For compact (U,fr)-manifolds of dimension 8, the above value 1/120 is not best possible. Denote by $\eta$ the Hopf symplectic line bundle over $S^4$. There is the disk bundle $D(\eta)$ with stable tangent bundle $p^!(\eta - 2) \varepsilon K(D(\eta))$, where $p : D(\eta) \longrightarrow S^4$ is projection. Now $D(\eta)/S(\eta) \approx HP(2)$, and there is Hopf symplectic line bundle $\eta'$ on $QP(2)$. It is easily seen that

$$\widetilde{K}(QP(2)) \approx K(D(\eta), S(\eta)) \longrightarrow \widetilde{K}(D(\eta)) \text{ maps } \eta' - 2$$

into $p^!(\eta - 2)$. Thus we may consider $D(\eta)$ as a compact (U,fr)-manifold with stable tangent bundle $\eta' - 2$. Then

$$\langle c_2^2(D(\eta)), \sigma(D(\eta))\rangle = 1$$

for an appropriate orientation, and all other Chern numbers of $D(\eta)$ are zero. Then

$$\mathrm{Td}[D(\eta)] = -3\, c_2^2[D(\eta)]/720 = -1/240.$$

The value 1/240 is best possible in dimension 8.

(15.1) THEOREM. Let $M^{2n}$ be a compact (U,fr)-manifold. A necessary and sufficient condition that there exist a closed U-manifold $N^{2n}$ with the same Chern numbers as $M^{2n}$ is that $\mathrm{Td}[M^{2n}]$ be an integer.

Proof. The necessity is clear. We prove the sufficiency. Suppose that $M^{2n}$ is a compact (U,fr)-manifold with $\mathrm{Td}[M^{2n}]$ an integer. Let

$$a_{j_1,\ldots,j_\chi} = c_{j_1}\ldots c_{j_\chi}[M^{2n}]$$

for each partition of n, and let

$$a = (a_{j_1,\ldots,j_\chi} : j_1 \geq \ldots \geq j_\chi, j_1 + \ldots + j_\chi = n).$$

We show that $\underline{s}_\omega(a)$ is an integer for each partition $\omega$ with $d(\omega) \leq n$.

There is $\tau = \tau(M, \partial M) \in K(M, M)$ and the Atiyah classes $\gamma_k(\tau) \in K(M, \partial M)$. Thus for $\omega > 0$ there are the classes $\underline{s}_\omega(\tau) \in K(M, \partial M)$. According to section 13,

$$\underline{s}_\omega[M] = \underline{s}_\omega(\tau)[M]$$

is then an integer for $\omega > 0$. But

$$\underline{s}_\omega(a) = \underline{s}_\omega[M],$$

hence $\underline{s}_\omega(a)$ is an integer for $\omega > 0$. But $\underline{s}_0(a) = Td[M^{2n}]$, which by assumption is an integer. The theorem now follows from (14.4).

16. The bordism groups $\Omega_*^{U,fr}$.

Suppose that $M^n$ is a compact (U,fr)-manifold. Embed $M^n$ smoothly in $I^{n+2k}$, $2k \geq n + 2$, so that $M \wedge \partial I^{n+2k} = \partial M$, so that this intersection is in a single face of $\partial I^{n+2k}$, and so that M is perpendicular to $\partial I^{n+2k}$ at this intersection. The normal bundle $\eta$ to M in $\partial I^{n+2k}$ may be supposed a complex vector space bundle with a given trivialization on $\partial M$ [12]. Let $\xi_k : E(\xi_k) \longrightarrow BU(k)$ be a universal U(k)-bundle, let $x_0 \in BU(k)$ and denote by F the fiber of $\xi_k$ above $x_0$. There is then a unique homotopy class of bundle maps

$$\begin{array}{ccc}(E(\eta),E(\eta)\,\partial M)) & \xrightarrow{f} & (E(\xi_k),F)\\ \downarrow & \bar{f} & \\ (M,\ \partial M) & \longrightarrow & (BU(k),x_0)\end{array}$$

Passing to disk bundles, we may consider f as a map

$$(D(\eta),S(\eta)\cup D(\eta|\,\partial M))\longrightarrow(D(\xi_k),S(\xi_k)\cup D^{2k})$$

and passing to quotients we get a map

$$g : D(\eta)/S(\eta)\cup D(\eta\,|\,\partial M)\longrightarrow MU(K)/S^{2k}.$$

There is the natural map

$$I^{n+2k}/\,\partial I^{n+2k}\longrightarrow D(\eta)/S(\eta)\cup D(\eta\,|\,\partial M)$$

collapsing $I^{n+2k}$ - Int $D(\eta)$ to a point. Composing these, we get from $M^n$ a map $S^{n+2k}\longrightarrow MU(k)/S^{2k}$ thus an element of

$$\pi_{n+2k}(MU(k)/S^{2k}),\ 2k \geq n + 2.$$

We shall in fact interpret $\pi_{n+2k}(MU(k)/S^{2k})$ as bordism classes of compact (U,fr)-manifolds. Thus we define

$$\Omega_n^{U,fr} = \pi_{n+2k}(MU(k)/S^{2k}),\ 2k \geq n + 2.$$

The method of Thom shows that every element of $\Omega_n^{U,fr}$ is represented by some compact (U,fr)-manifold. We could give a complete bordism description of this group, but we forego the tedious details.

Given an $M^{2n}$, $n > 0$, and the associated map $g : S^{2n+2k}\longrightarrow MU(k)/S^{2k}$, there is

$$\begin{array}{ccc} H^{2n}(BU(k)) & & H^{2n+2k}(S^{2n+2k}) \\ \simeq \big\downarrow \varphi & & \big\uparrow g^* \\ H^{2n+2k}(MU(k)) & \xleftarrow{q^*} & H^{2n+2k}(MU(k)/S^{2k}). \end{array}$$

The image $g^* q^{*-1} \varphi(c_{i_1} \cdots c_{i_\chi})$, $i_1 + \cdots + i_\chi = n$, are invariants of the homotopy class of g. But it can be seen that these can be considered as normal Chern numbers, namely

$$\langle g^* q^{*-1} \varphi(c_{i_1} \cdots c_{i_\chi}), \sigma(S^{2n+2k}) \rangle = \langle c_{i_1}(\eta) \cdots c_{i_\chi}(\eta), \sigma(M) \rangle$$

where $\tau + \eta = 0$. Since the Chern numbers $c_{j_1} \cdots _{j_r}[M]$ can be expressed in terms of the normal Chern numbers, it follows that they are bordism invariants. Hence so also is $Td[M^{2n}]$ a bordism invariant. Thus we may consider Td a homomorphism

$$Td : \Omega^{U,fr}_{2n} \longrightarrow Q.$$

The following interpretation of Td we owe to P. S. Landweber.

(16.1) <u>Let</u> $M^{2n}$, $n > 0$, <u>be a compact</u> (U,fr)-<u>manifold and let</u> $f : S^{2n+2k} \longrightarrow MU(k)/S^{2k}$ <u>be an associated map</u>. <u>Denote by</u> $T^{(k)}$ <u>the Thom class in</u> $\tilde{K}(MU(k))$, <u>let</u> $\sigma$ <u>be the orientation class of</u> $S^{2n+2k}$ <u>and consider</u>

$$H_{2n+2k}(S^{2n+2k}) \xrightarrow{f_*} H_{2n+2k}(MU(k)/S^{2k}) \underset{\approx}{\xleftarrow{q_*}} H_{2n+2k}(MU(n)).$$

<u>Then</u>

$$Td[M^{2n}] = \langle ch\ T^{(k)}, q_*^{-1} f_*(\sigma) \rangle .$$

Proof. If $M^{2n}$ is a closed U-manifold with associated map $g : S^{2n+2k} \longrightarrow MU(k)$, it may be verified that

$$T[M^{2n}] = \langle ch\ T^{(k)}, g_*(\sigma) \rangle .$$

Define for any compact (U,fr)-manifold $M^{2n}$,

$$T'[M^{2n}] = \langle ch\ T^{(k)}, q_*^{-1} f_*(\sigma) \rangle .$$

Consider $T,T' : \Omega_{2n}^{U,fr} \longrightarrow Q$. It follows that $T = T'$ on the image of $u : \Omega_{2n}^{U} \longrightarrow \Omega_{2n}^{U,fr}$. Since $\Omega_{2n}^{U,fr}$/Image u is finite, we must have $T = T'$ in all cases. The remark follows.

In the diagram

$$\begin{array}{ccccccc} 0 \longrightarrow \Omega_{2n}^{U} & \xrightarrow{u} & \Omega_{2n}^{U,fr} & \xrightarrow{\partial} & \Omega_{2n-1}^{fr} & \longrightarrow 0 \\ & & \downarrow{\scriptstyle Td} & & & \\ & & Q & & & \end{array}$$

note that the image of (Td)u is the integers, hence we get a homomorphism

$$\Omega_{2n-1}^{fr} \approx \Omega_{2n}^{U,fr}/\text{Image } u \xrightarrow{Td} Q/Z.$$

Denote by E the composite homomorphism

$$E : \Omega_{2n-1}^{fr} \longrightarrow Q/Z.$$

Recall now the homomorphism

$$e_c : \{S^{2n-1}, S^0\} \longrightarrow Q/Z$$

of J. F. Adams [3]. Namely let

$$f : S^{2n+2k-1} \longrightarrow S^{2k}$$

represent an element $\alpha$ of $\{S^{2n-1}, S^0\}$. Attach a (2n + 2k)-cell to $S^{2k}$

via f, thus obtaining a space X. Denote by

$$\lambda_{2k} \in H^{2k}(X),\ \lambda_{2n+2k} \in H^{2n+2k}(X)$$

generators induced by the standard orientations of the spheres. According to Atiyah-Hirzebruch [6], there exists $x \in \widetilde{K}(X)$ with

$$\mathrm{ch}\ x = \lambda_{2n} + r\ \lambda_{2n+2k},\ r \text{ rational.}$$

Then r mod 1 ε Q/Z is a function only of $\alpha$ and Adams defines

$$e_c(\alpha) = r \bmod 1 \in Q/Z.$$

We give a proof due to P. S. Landweber of the following theorem; it replaces a more awkward proof of our own.

(16.2) THEOREM. Using the natural identifications

$$\Omega^{fr}_{2n-1} \simeq \pi_{2n+2k-1}(S^{2k}) \approx \{S^{2n-1}, S^o\}$$

for 2k large, the homomorphism $E : \Omega^{fr}_{2n-1} \longrightarrow Q/Z$ coincides with the homomorphism $e_c : \{S^{2n-1}, S^o\} \longrightarrow Q/Z$ of Adams.

Proof. Suppose given a map $f : S^{2n+2k-1} \longrightarrow S^{2k}$ representing an element of $\Omega^{fr}_{2n-1}$. Using the natural embedding $i : S^{2k} \subset MU(k)$, we get $f' = if : S^{2n+2k-1} \longrightarrow MU(k)$. Regard f' as representing an element of $\pi_{2n+2k-1}(MU(k)) = \Omega^U_{2n-1} = 0$, there exists an extension of f' to a map

$$(D^{2n+2k}, S^{2n+2k-1}) \longrightarrow (MU(k), S^{2k}),$$

from which, passing to quotients, we get a map

$$g : S^{2n+2k} \longrightarrow MU(k)/S^{2k},$$

representing an element $\beta$ of $\Omega^{U,fr}_{2n}$. Moreover $\partial : \Omega^{U,fr}_{2n} \longrightarrow \Omega^{fr}_{2n-1}$

clearly maps $\beta$ into the given element $\alpha$ of $\Omega^{fr}_{2n-1}$. Hence $E(\alpha) = \mathrm{Td}\,\beta \bmod 1$. By (16.1) we have

$$\mathrm{Td}\,\beta = \langle \mathrm{ch}\, T^k, q_*^{-1} g_*(\sigma) \rangle .$$

Recall now the space X obtained by attaching $D^{2n+2k}$ to $S^{2k}$ via $f : S^{2n+2k-1} \longrightarrow S^{2k}$. The above map

$$(D^{2n+2k}, S^{2n+2k-1}) \longrightarrow (MU(k), S^{2k})$$

gives rise to a map $h : X \longrightarrow MU(k)$ so that

$$\begin{array}{ccc} X & \xrightarrow{h} & MU(k) \\ \big\downarrow p & & \big\downarrow q \\ S^{2n+2k} & \xrightarrow{g} & MU(k)/S^{2k} \end{array}$$

is commutative. Here p is the collapsing map $X \longrightarrow S^{2n+2k} = D^{2n+2k}/S^{2n+2k-1}$. From

$$\begin{array}{ccccc} S^{2k} & \longrightarrow & X & \longrightarrow & S^{2n+2k} \\ \big\downarrow = & & \big\downarrow h & & \big\downarrow g \\ S^{2k} & \longrightarrow & MU(k) & \longrightarrow & MU(k)/S^{2k} \end{array}$$

we see that $h^* : H^*(MU(k)) \longrightarrow H^*(X)$ maps the Thom class of $H^{2k}(MU(k)$ into $\wedge_{2k}$. The Thom class is the lead term of ch $T^k$, hence

$$\mathrm{ch}\, h^!(T^k) = \wedge_{2k} + r \wedge_{2n+2k}$$

hence we may choose $x = h^!(T^k)$ in Adams definition. Let $\sigma' \in H_{2n+2k}(X)$ have $\langle \wedge_{2n+2k}, \sigma' \rangle = 1$. Then

$$
\begin{aligned}
E(\alpha) &= \langle \mathrm{ch}\, T^{(k)}, q_*^{-1} g_*(\sigma) \rangle \mod 1 \\
&= \langle \mathrm{ch}\, T^{(k)}, h_*(\sigma') \rangle \mod 1 \\
&= \langle h^* \mathrm{ch}\, T^{(k)}, \sigma' \rangle \mod 1 \\
&= r \mod 1 \\
&= e_c(\alpha).
\end{aligned}
$$

The theorem follows.

We are now in a position to borrow the results of Adams [3] which completely analyze the image of $e_c$. For each positive integer n, denote by $B_n$ the nth Bernoulli number. Denote by $a_n$ the denominator of $B_n/4n$ in lowest terms (for references, see [2,20]). Let

$$d_{2n} = a_{2n}, d_{2n+1} = a_{2n+1}/2.$$

According to Adams [3], the image of $e_c : \{S^{2n-1}, S^o\} \longrightarrow Q/Z$ consists precisely of the integral multiples of the following numbers;

(a) $\{S^{8k-1}, S^o\}$, multiples of $1/d_{2k}$

(b) $\{S^{8k-3}, S^o\}$, multiples of 1

(c) $\{S^{8k-5}, S^o\}$, multiples of $1/d_{2k-1}$

(d) $\{S^{8k-7}, S^o\}$, multiples of 1/2.

(16.3) COROLLARY. _The homomorphism_ Td : $\Omega_{2n}^{U,fr} \longrightarrow Q$ _maps precisely onto the integral multiples of the following numbers:_

$\Omega_{8k}^{U,fr}$ _onto multiples of_ $1/d_{2k}$

$\Omega_{8k-2}^{U,fr}$ _onto multiples of_ 1

$\Omega_{8k-4}^{U,fr}$ _onto multiples of_ $1/d_{2k-1}$

$\Omega_{8k-6}^{U,fr}$ _onto multiples of_ 1/2.

We can now answer completely the question of section 15. In

$$0 \longrightarrow \Omega^{U}_{2n} \xrightarrow{u} \Omega^{U,fr}_{2n} \xrightarrow{\partial} \Omega^{fr}_{2n-1} \longrightarrow 0,$$

define $D \subset \Omega^{fr}_{2n-1}$ to be the image under $\partial$ of Tor $\Omega^{U,fr}$. It is easily verified that $\alpha \in D$ if and only if given $[M^{2n}] \in \Omega^{U,fr}_{2n}$ with $\partial[M^{2n}] = \alpha$ there exists a closed U-manifold having the same Chern numbers as $M^{2n}$. Now $\partial$ maps $\Omega^{U,fr}_{2n}$/(Image u + Tor $\Omega^{U,fr}_{2n}$) isomorphically onto $\Omega^{fr}_{2n-1}/D$. But by (16.3) and (15.1), $\Omega^{fr}_{2n-1}/D$ is then a cyclic group; in fact

$$\Omega^{fr}_{8k-1}/D \approx Z_{d_{2k}}, \quad \Omega^{fr}_{8k-3}/D = 0,$$

$$\Omega^{fr}_{8k-5}/D \approx Z_{d_{2k-1}}, \quad \Omega^{fr}_{8k-7}/D \approx Z_2.$$

Put negatively, an element of $\Omega^{fr}_{2n-1}$ can be detected by Chern numbers if and only if it can be detected by the Adams homomorphism $e_c$.

Note also that $\partial$ maps Tor $\Omega^{U,fr}_{2n}$ isomorphically onto D. This can be used to give Tor $\Omega^{U,fr}_{2n}$ in low dimensions:

| n | 2 | 3 | 4 | 5 | 6 | 7 |
|---|---|---|---|---|---|---|
| D | 0 | $Z_2$ | $Z_2$ | 0 | 0 | $Z_2$ |

(see Toda [25]).

We can also define groups $\Omega^{SU,fr}_{n}$. It is possible to define them as all bordism classes of compact (SU,fr)-manifold. However we define them here by

$$\Omega^{SU,fr}_{n} = \pi_{n+2k}(MSU(k)/S^{2k}), \text{ k large.}$$

There is an exact sequence

$$\cdots \longrightarrow \Omega^{SU}_{n} \longrightarrow \Omega^{SU,fr}_{n} \longrightarrow \Omega^{fr}_{n-1} \longrightarrow \Omega^{SU}_{n-1} \longrightarrow \cdots.$$

Also there are homomorphisms Td : $\Omega^{SU}_{2n} \longrightarrow Z$ and Td : $\Omega^{SU,fr}_{2n} \to Q$. Generalizations of the following are very well-known.

(16.4) Let $M^{8k+4}$ be a closed SU-manifold. Then Td$[M^{8k+4}]$ is even.

Proof. The manifold $M^{8k+4}$ gives rise to an associated map

$$g : S^{8n+8k} \longrightarrow MSU(4k - 2)$$

and

$$\begin{aligned} \mathrm{Td}[M^{8k+4}] &= \langle \mathrm{ch}\, T^{(4n-2)}, g_*(\sigma) \rangle \\ &= \langle \mathrm{ch}\, g^! T^{(4n-2)}, \sigma \rangle. \end{aligned}$$

However it follows from Chapter I that $T^{(4n-2)}$ is symplectic, hence $g^! T^{(4n-2)}$ is in the image of

$$\widetilde{KSp}(S^{8n+8k}) \longrightarrow \widetilde{K}(S^{8n+8k})$$

which maps a generator onto twice a generator. Hence Td$[M^{8k+4}]$ is even.

Since $\Omega^{SU}_{8k+3} = 0$ and $\Omega^{SU}_{8k+5} = 0$ [12], we have the diagram

$$\begin{array}{ccccccccc} 0 & \longrightarrow & \Omega^{SU}_{8k+4} & \xrightarrow{u} & \Omega^{SU,fr}_{8k+4} & \longrightarrow & \Omega^{fr}_{8k+3} & \longrightarrow & 0 \\ & & & & \Big\downarrow {\scriptstyle (1/2)\mathrm{Td}} & & & & \\ & & & & Q & & & & \end{array}$$

and we see from (16.4) that (1/2)Td maps image u into Z. Hence we get an induced homomorphism

$$E_{SU} : \Omega^{fr}_{8k+3} \longrightarrow Q/Z$$

which coincides with the homomorphism $e_R$ of Adams [3]. Let $E_{SU} = E_U$ on $\Omega^{fr}_{8k-1}$.

(16.5) The homomorphism $E_{SU} : \Omega^{fr}_{4k-1} \longrightarrow Q/Z$ maps onto all integral multiples of $1/a_k$ where $a_k$ is the denominator of $B_k/4k$ in lowest terms.

## 17. The groups $\Omega^{U,SU}_*$.

In these last two sections we shall outline within the framework of these notes a complete proof of the assertions that $e_c : \Omega^{fr}_{8k+5} \longrightarrow Q/Z$ is trivial and that the image of $e_c : \Omega^{fr}_{8k+1} \longrightarrow Q/Z$ is $Z_2$. In particular the argument for the second part will complete our proof of the Anderson-Brown-Peterson theorem discussed in section 11.

By analogy with $\Omega^{U,fr}_n$ we shall geometrically define the bordism groups $\Omega^{U,SU}_n$. Let $(2m - n)R \oplus \tau \longrightarrow B^n$ be the stable tangent bundle of a compact manifold with boundary. Let $\mathcal{U} \longrightarrow B^n$ and $\mathcal{SU} \longrightarrow B^n$ respectively denote the associated bundle with fibre $O(2m)/U(m)$ and $O(2m)/SU(m)$. There is the principal fibring $\mathcal{SU} \longrightarrow \mathcal{U}$ with fibre $U(m)/SU(m) = U(1)$. A (U,SU)-structure on $B^n$ is a pair consisting of a homotopy class of cross-sections of $\mathcal{SU} \longrightarrow B^n$ defined over $\partial B^n$ together with a compatible class of cross-sections of $\mathcal{U} \longrightarrow B^n$ defined over all of $B^n$. This is independent of m for m large [12,(2.3)]. Such a (U,SU)-structure induces a natural SU-structure on $\partial B^n$ of course. Note that the cross-section of $\mathcal{U} \longrightarrow B^n$ induces a principal U(1)-bundle over $B^n$, which along $\partial B^n$ already has a homotopy class of cross-sections; thus, the first Chern class $c_1(B^n)$ lies in $H^*(B^n, \partial B^n; Z)$. The remaining Chern classes lie in $H^*(B^n; Z)$ of course.

We shall say $B^n$ bords if and only if $B^n \subset \partial W^{n+1}$ as a compact regular submanifold where

1) $V^n = \partial W^{n+1} \smallsetminus (B^n)^o$ admits an SU-structure extending that on $\partial B^n = \partial V^n$

ii) $W^{n+1}$ admits a U-structure compatible with that on $\partial W^{n+1}$.

We should observe that $c_1(W^{n+1}) \in H^2(W^{n+1}, V^n; Z)$ and under the induced homomorphism $H^2(W^{n+1}, V^n; Z) \longrightarrow H^2(B^n, \partial B^n; Z), c_1(W^{n+1}) \longrightarrow c_1(B^n)$. We can define $-B^n$ suitably and in the usual way arrive at the bordism groups $\Omega_n^{U,SU}$ together with an exact sequence

$$\cdots \longrightarrow \Omega_n^{SU} \longrightarrow \Omega_n^{U} \longrightarrow \Omega_n^{U,SU} \longrightarrow \Omega_{n-1}^{SU} \longrightarrow \cdots$$

If $n = 2 \pmod 4$ we can define $Td : \Omega_n^{U,SU} \longrightarrow Q$. We recall [16] that for $n/2$ odd, $T_{n/2}(c_1, \ldots, c_{n/2}) = c_1 P_{n/2}(c_1, \ldots, c_{n/2})$. Since $c_1 \in H^2(B^n, \partial B^n; Z)$ and $P_{n/2}(c_1, \ldots, c_{n/2}) \in H^{n-2}(B^n; Q)$ we have $T_{n/2}(c_1, \ldots, c_{n/2}) \in H^n(B^n, \partial B^n; Z)$ so we can put
$Td[B^n] = \langle T_{n/2}(c_1, \ldots, c_{n/2}), \sigma_{2n} \rangle \in Q$. Suppose $B^n$ bords, then $B^n \subset W^{n+1}$ as described, and
$c_1 \in H^2(W^{n+1}, V^n; Z), P_{n/2}(c_1, \cdots, c_{n/2}) \in H^{n-2}(W^{n+1}; Z)$, hence $T_{n/2}(c_1, \ldots, c_{n/2}) \in H^n(W^{n+1}, V^n; Z)$. On the other hand, the fundamental class generates the kernel of $H_n(B^n, \partial B^n; Z) \longrightarrow H_n(W^{n+1}, V^n; Z)$, so $Td(B^n) = 0$ by the usual reasoning. This shows $Td : \Omega_n^{U,SU} \longrightarrow Q$ is well defined if $n = 2 \pmod 4$.

(17.1) LEMMA: *If* $n = 6 \pmod 8$ *then* $\Omega_n^{U,SU} \longrightarrow Q/Z$ *is trivial, but if* $n = 2 \pmod 8$ *the image is* $Z_2$.

We showed in [12,(18.3)] that $\Omega_{8k+5}^{SU} = 0$, hence we have $\Omega_{8k+6}^{U} \longrightarrow \Omega_{8k+6}^{U,SU} \longrightarrow 0$, thus $Td : \Omega_{8k+6}^{U,SU} \longrightarrow Q$ is integral valued. In the second case, $\Omega_{8k+1}^{SU}$ consists entirely of elements of order 2, hence we see immediately that $2\,Td : \Omega_{8k+1}^{U,SU} \longrightarrow Q$ is integral valued.

Finally we must see that the value 1/2 is taken on for $8k + 2$. The tangent bundle of the closed 2-cell

$$D^2 = \left\{ z \,\middle|\, |z| \leq 1 \right\}$$

may be identified with $D^2 \times C$. Along $\partial D^2 = S^1$ there is a non-zero cross-section $z \to (z,z)$. Thus $D^2$ becomes a (U,SU)-manifold. It is well known that $c_1(D^2)$, the obstruction to the extension of this particular cross-section, is the generator of $H^2(D^2, \partial D^2, Z)$, hence Td $[D^2] = \overset{+}{-} 1/2$. In general, let $M^{8k}$ be a closed SU-manifold. Since $\partial(D^2 \times M^{8k}) = S^1 \times M^{8k}$ this product is naturally a (U,SU)-manifold. The reader can show Td$[D^2 \times M^{8k}] = 1/2$ Td$[M^{8k}]$. We take Td$[M^{8k}] = 1$ to see that the value $1/2$ is taken on in dimension $8k + 2$.

If we also define $\Omega_n^{SU,fr}$ we obtain a commutative diagram

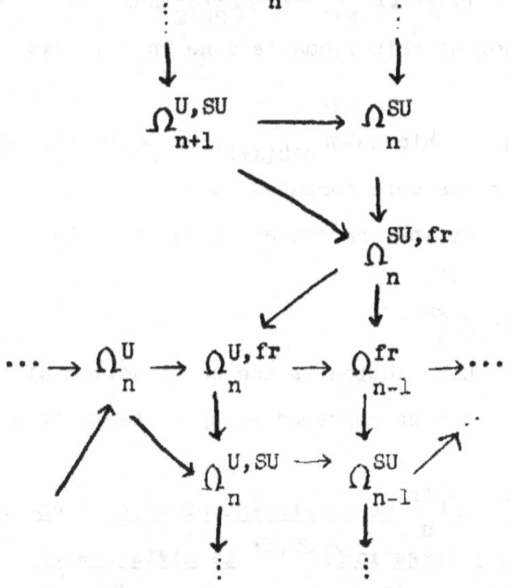

where the two vertical sequences and the two hroizontal are both exact. Note that for $n \equiv 2 \bmod 4$

$$\begin{array}{c} \Omega_n^{U,fr} \\ \downarrow \\ \Omega_n^{U,SU} \end{array} \overset{Td}{\underset{Td}{\rightrightarrows}} Q$$

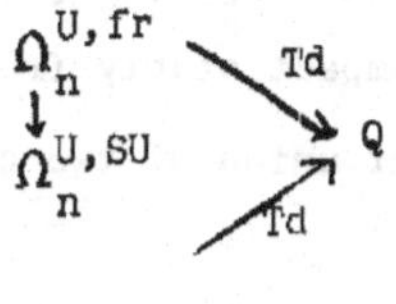

is commutative. From this diagram we see that if $\alpha \in \ker(\Omega^{fr}_{n-1} \to \Omega^{SU}_{n-1})$ then $e_c(\alpha) = 0 \in Q/Z$, but if $n = 6 \bmod (8)$ $\Omega^{SU}_{8k+5} = 0$, thus $e_c : \Omega^{fr}_{8k+5} \to Q/Z$ is trivial. On the other hand $2\alpha$ lies in the kernel for $8k+1$, hence $e_c : \Omega^{fr}_{8k+1} \to Q/Z$ has image at most $Z_2$. To show this image is exactly $Z_2$ we need

(17.2) LEMMA: <u>There is an element</u> $[M^{8k}] \in \Omega^{SU}_{8k}$ <u>for which</u> $Td[M^{8k}] = 1$, <u>and for which</u> $[M^{8k}][S^{-1}]$ <u>is in the image of</u> $\Omega^{fr}_{8k+1} \to \Omega^{SU}_{8k+1}$.

If $\alpha \in \Omega^{fr}_{8k+1}$ is the element, then $e_c(\alpha) \neq 0$ since if $\beta \in \Omega^{U,fr}_{8k+2}$ has $\partial\beta = \alpha$ then the image of $\beta$ in $\Omega^{U,SU}_{8k+2}$ differs from $[M^{8k}][D^2]$ by an element of the image $\Omega^{U}_{8k+2} \to \Omega^{U,SU}_{8k+2}$ and $Td([M^{8k}][D^2]) = 1/2$. The proof of this lemma is done in the next section.

The groups $\Omega^{U,SU}_n$ are isomorphic to $\pi_{n+2(k+1)}(CP(\infty)/CP(1) \wedge MSU(k))$. In [12,(14.5)] the homotopy groups were computed, and $\Omega^{U,SU}_n \simeq \Omega^{SU}_{n-2} + \Omega^{U}_{n-4}$. In fact the present exact sequence involving $\Omega^{U,SU}_*$ is the same as [12,(15.1)].

## 18. The image of $\Omega^{fr}_*$ in $\Omega^{SU}_*$

The purpose of this section is the construction of those elements in $\Omega^{SU}_{8k+1}$ which can be represented by a stably framed closed manifold.

(18.1) LEMMA: <u>Let</u> $[V^n] \in \Omega^{fr}_n$ <u>be an element of order</u> 2 <u>and let</u> $[M^k] \in \Omega^{SU}_k$ <u>be an element whose image in</u> $\Omega^{SU,fr}_k$ <u>is divisible by</u> 2. <u>There is then an element</u> $[M^{n+k}] \in \Omega^{fr}_{n+k}$ <u>whose image in</u> $\Omega^{SU}_{n+k}$ <u>is</u> $[M^k][V^n]$ <u>and</u> $2[M^{n+k}] = 0 \in \Omega^{fr}_{n+k}$.

Proof: Let $(B^k, \Phi, \varphi)$ be such that $2[B^k] = [M^k]$ in $\Omega^{SU,fr}_k$ then $2[\partial B^k] = 0$ in $\Omega^{fr}_{k-1}$. There is a compact stably framed manifold $C^k$ whose boundary, $\partial C^k$, is the disjoint union of two copies of $\partial B^k$, labeled $\partial B_1^k$ and $\partial B_2^k$. There is also a compact stably framed manifold $C^{n+1}$ whose boundary is the disjoint union of two copies of $V^n$, labeled $V_1^n$ and $V_2^n$.

We consider $C_1^{n+k} = (C^k \times V_1^n \cup (-1)^k \partial B_1^k \times C^{n+1})$. Since $\partial(C^k \times V_1^n) = \partial B_1^k \times V_1^n \cup B_2^k \times V_1^n$ and $(-1)^k \partial(\partial B_1^k \times C^{n+1}) = -\partial B_1^k \times V_1^n \cup -\partial B_1^k \times V_2^n$ we see that $C^{n+k}$ is a compact stably framed manifold with

$$\partial C_1^{n+k} = (\partial B_2^k \times V_1^n) \cup -(\partial B_1^k \times V_2^n).$$

The two ends of $C_1^{n+1}$ can be identified to form a closed stably framed manifold $M^{n+k}$. We also have

$$C_2^{n+k} = (C^k \times V_2^n \cup (-1)^k \partial B_2^k \times C^{n+1})$$

and

$$\partial C_2^{n+k} = (\partial B_1^k \times V_2^n) \cup -(\partial B_2^k \times V_1^n)$$

Of course $C_1^{n+k}$ and $C_2^{n+k}$ are diffeomorphic as stably framed manifolds, thus the closed stably framed manifold $C_1^{n+k} \cup C_2^{n+k}$ represents $2[M^{n+k}]$. Observe, however, that $(-1)^k \partial(C^k \times C^{n+1}) = (-1)^k[\partial B_1^k \times C^{n+1} \cup (-1)^k(C^k \times V_1^n)] \cup [(\partial B_2^k \times C^{n+1} \cup (-1)^k C^k \times V_2^n] = C_1^{n+k} \cup C_2^{n+k}$, so $2[M^{n+k}] = 0$ in $\Omega_{n+k}^{fr}$.

To see that the image of $[M^{n+k}]$ in $\Omega_{n+k}^{SU}$ is $[M^k][V^n]$ we first note that

$$(-1)^k \partial(B_1^k \times C^{n+1}) = (-1)^k(\partial B_1^k \times C^{n+1}) \cup (B_1^k \times V_1^n \cup B_1^k \times V_2^n)$$

Since $2[B^k] = [M^k]$ in $\Omega_k^{SU,fr}$ there is a $W^{k+1}$ with $M^k \cup (-B_1^k) \cup (-B_2^k)$ in $\partial W^{k+1}$. There is no loss of generality in assuming $\partial W^{k+1} = M^k \cup (-B_1^k) \cup (-B_2^k) \cup C^k$. We now form $W^{k+1} \times V_1^n$ so

$$\partial(W^{k+1} \times V_1^n) = M^n \times V_1^n \cup -(B_1^k \times V_1^n) \cup -(B_2^k \times V_1^n) \cup -(C^k \times V_1^n).$$

We glue $(-1)^k B^k \times C^{n+1}$ to $W^{k+1} \times V_1^n$ along $B_1^k \times V_1^n \cup B_2^k \times V_1^n$. The boundary of the result is a disjoint copy of $M^k \times V_1^n$ together with a copy of $M^{n+k}$, that is, of $-(C^k \times V_1^n \cup \partial B_1^k \times V_1^n \cup \partial B_1^k \times V_2^n)$ with $\partial B_2^k \times V_1^n = \partial B_1^k \times V_2^n$ hence $[M^{n+k}] = [M^k][V^n]$ in $\Omega_{n+k}^{SU}$.

(18.2) THEOREM. *There is an element* $[M^8] \varepsilon \Omega_8^{SU}$ *with Todd genus* 1 *for which* $[M^8]^n \times [\bar{S}^1] \neq 0$ *lies in the image* $\Omega_{8n+1}^{fr} \longrightarrow \Omega_{8n+1}^{SU}$.

Proof: We first see that $[M^8]^n \times [\bar{S}^1] \neq 0$ by (11.1) since $[M^8]$ is to have odd Todd genus. In section 15 we showed that $\Omega_8^{SU,fr}/\text{im}(\Omega_8^{SU}) \simeq Z_{240}$ and we constructed a bordism class in $\Omega_8^{SU,fr}$ with Todd genus 1/240. There is, then a $[B^8] \varepsilon \Omega^{SU,fr}_8$ with $Td[B^8] = 1/2$, and $[M^8] \varepsilon \Omega_8^{SU}$ with $[M^8] \longrightarrow 2[B^8] \varepsilon \Omega_8^{SU,fr}$.

We prove (18.2) for this bordism class by induction. It is valid for n = 0. Suppose there is $[V^{8n+1}] \varepsilon \Omega_{8n+1}^{fr}$, with $2[V^{8n+1}] = 0$, whose image is $[M^8]^n \times [\bar{S}^1] \varepsilon \Omega_{8n+1}^{SU}$. We apply (18.1) with $[V^n] = [V^{8n+1}]$ and $[M^k] = [M^8]$ to obtain $[V^{8(n+1)+1}]$.

This provides a full proof of the Anderson-Brown-Peterson theorem of section 11. Geometric constructions used in (18.2) are a paraphrase of the Toda bracket formation used by Adams.

## BIBLIOGRAPHY


1. J. F. Adams, Cohomology operations, Seattle Conference on Differential and Algebraic Topology, A.M.S. 1962.

2. _________, On the groups J(X). II.

3. _________, On the groups J(X). IV.

4. D. Anderson, E. H. Brown and F. Peterson, SU-bordism, KO-characteristic numbers, and the Kervaire invariant, Ann. of Math.

5. M. F. Atiyah, Immersions and embeddings of manifolds, Topology 1 (1962), 125-132.

6. M. F. Atiyah and F. Hirzebruch, Vector bundles and homogeneous spaces, Proc. Symp. Pure Math., 3. Differential Geometry, Amer. Math. Soc. (1961), 7-38.

7. M. F. Atiyah, R. Bott, and A. Shapiro, Clifford modules, Topology 3 (1964), Suppl. 1, 3-38.

8. A. Borel, Sur la cohomolgie des espaces fibres principaux et des espaces homogènes des groupes de Lie compact. Ann. Math. 57, 115-207 (1953).

9. R. Bott, An application of the Morse theory to the topology of Lie groups, Bull. Soc. Math. France, 84 (1956), 251-281.

10. P. E. Conner and E. E. Floyd, Differentiable Periodic Maps, Springer-Verlag, 1964.

11. _________, Fixed point free involutions and equivariant maps. Bull. Amer. Math. Soc. 66, 416-441 (1960).

12. _________, Torsion in SU-Bordism, Mem. A.M.S. 60 (1966).

13. A. Dold, Relations between ordinary and extraordinary cohomology, Colloquium on Algebraic Topology, Aarhus (1962).

14. E. Dyer, Chern characters of certain complexes, Math. Z. 80 (1963), 363-373.

15. A. Hattori, Integral characteristic numbers for weakly complex manifolds, Univ. of Tokyo (mimeo. notes 1966).

16. F. Hirzebruch, Neue topologische Methoden in der algebraischen Geometrie, Springer-Verlag, 1956.

17. R. Lashof, Poincare duality and cobordism, Amer. J. of Math. 109, 257-277 (1963).

18. A. Liulevicious, Notes on homotopy of Thom spectra, Amer. J. Math. 86, 1 - 16 (1964).

19. J. W. Milnor, On the cobordism ring $\Omega^*$ and a complex analogue, Amer. J. Math. 82, 505-521 (1960).

20. J. W. Milnor and M. A. Kervaire, Bernoulli numbers, homotopy groups, and a theorem of Rohlin, Proc. Int. Cong. Math. 1958, Cambridge (1960).

21. S. P. Novikov, Homotopy properties of Thom complexes, Mat. Sb. 57 (1962), 407-442 (Russian).

22. R. S. Palais et al, Seminar on the Atiyah-Singer Index Theorem, Ann. of Math. Studies 57 (1965), Princeton.

23. R. E. Stong, Relations among characteristic number - I, Topology 4 (1965), 267-281.

24. R. Thom, Travaux de Milnor sur le cobordisme, Bourbaki Seminar notes 1958-59, Paris.

25. H. Toda, Composition Methods in Homotopy Groups of Spheres, Ann. of Math. Studies 49, 1962.

26. G. W. Whitehead, Generalized homology theories, Trans. Amer. Math. Soc., 102, 227-283 (1962.

Offsetdruck: Julius Beltz, Weinheim/Bergstr.